Multicriteria Evaluation in a Fuzzy Environment

Contributions to Economics

Giuseppe Munda

Multicriteria Evaluation in a Fuzzy Environment

Theory and Applications in Ecological Economics

With 28 Figures

Physica-Verlag

A Springer-Verlag Company

Series Editors
Werner A. Müller
Peter Schuster

Author
Giuseppe Munda Ph. D.
Free University Amsterdam
Department of Economics
Institute of Regional Economics
De Boelelaan 1105
NL-1081 HV Amsterdam, The Netherlands

ISBN 978-3-7908-0892-6   ISBN 978-3-642-49997-5 (eBook)
DOI 10.1007/978-3-642-49997-5

Die Deutsche Bibliothek - CIP-Einheitsaufnahme
Munda. Giuseppe:
Multicriteria evaluation in a fuzzy environment: theory and
applications in ecological economics / Giuseppe Munda. –
Heidelberg : Physica-Verl., 1995
(Contributions to economics)

to my parents

PREFACE

This book is the result of some years of research carried out at the Vrije Universiteit of Amsterdam and at the Joint Research Centre of the European Commission.

The awareness of actual and potential conflicts between economic progress in production, consumption, and technology and the environment has led to the concept of *"sustainable development"*, implying that economic and ecological values are well balanced in evaluation and decision making.

The linkages between ecosystems and economic systems are the focus of ecological economics. In ecological economics, a *multidimensional* approach to economic and policy-making is emphasised. In this book, the introduction of multicriteria decision aid techniques in the framework of ecological economics is widely discussed. Since such techniques are based on a "constructive" rationality and allow one to take into account conflictual, multidimensional, incommensurable and uncertain effects of decisions, they can be considered perfectly consistent with the methodological foundations of ecological economics.

Since here the assumption is accepted that efficiency, equity and sustainability are the three conflictual values of economics, a mathematical procedure able to deal with these issues in an operational framework is developed, with a particular view on imprecise information in a practical environmental planning context.

Given the problem of the differences in the measurement levels of the variables used for economic-ecological modelling, multicriteria methods able to deal with mixed information (both qualitative and quantitative measurements) can be considered particularly useful. Another problem related to the available information concerns the uncertainty (stochastic and/or fuzzy) contained in this information. Fuzzy uncertainty does not concern the occurrence of an event but the event itself, in the sense that it cannot be described unambiguously. This situation is very common in human systems. Spatial-environmental systems in particular, are complex systems characterised by subjectivity, incompleteness and imprecision. Therefore, the combination of different levels of measurement with different types of uncertainty has to be considered as an important research issue in multicriteria evaluation.

Our overview of multicriteria evaluation methods in a fuzzy environment, shows that traditional fuzzy multicriteria methods are utility based models; most of these methods are limited to the use of only fuzzy information (often only triangular fuzzy numbers) and a key issue is how to compare fuzzy sets. Thus a new approach based on a semantic distance using areas instead of traditional intersections is presented. This new semantic distance overcomes different weak points of traditional comparison methods. A new multicriteria method, based on some aspects of the partial comparability axiom, called NAIADE (**N**ovel **A**pproach to **I**mprecise **A**ssessment and **D**ecision **E**nvironments) is developed. It is a discrete multicriteria method whose impact (or evaluation) matrix may include either crisp, stochastic or fuzzy measurements of the performance of an alternative a_n with respect to a judgement criterion g_m, thus it is very flexible for real-world applications.

Since in environmental and resource management and policy aiming at an ecologically sustainable development many conflicting issues and interests emerge, particular attention has to be given to the problem of different values and goals of different groups in society. Equity and conflicting values in multicriteria decision aid are traditionally introduced in two different ways:

(1)   by weighting the different criteria, but often in public decision making a single point-value solution (e.g. weights) tends to lead to deadlocks in a decision process because it imposes too rigid conditions to reach a compromise;

(2)   by taking into consideration a set of ethical evaluation criteria. A weak point of this approach is that it could lead to an excessive number of evaluation criteria. Furthermore, to identify ethical criteria may be not an easy task.

We therefore propose a third possibility i.e. the use of conflict analysis procedures to be integrated with multicriteria evaluation in order to allow policy-makers to seek for "defendable" decisions that could reduce the degree of conflict (in order to reach a certain degree of consensus) or that could have a higher degree of equity on different income groups. Therefore, finally a compromise solution taking into account all the three conflictual values of economics (efficiency, equity and sustainability) can in principle be identified.

The empirical relevance of the proposed mathematical procedures will be illustrated by means of a real-world environmental management problem (in the area of the Delta of the river Po in Italy).

ACKNOWLEDGEMENTS

I am particularly grateful to three persons who devoted a lot of time and energies to support the research whose results are presented in this book: Peter Nijkamp and Piet Rietveld at the Vrije Universiteit and Massimo Paruccini at the Joint Research Centre.

I am indebted to Robert Costanza, Sylvie Faucheux, Silvio Funtowicz, Joan Martinez-Alier, Martin O'Connor and Jerome Ravetz for having let me discover the fascinating field of ecological economics.

The helpful comments and suggestions of Benedetto Matarazzo and Jaap Spronk on previous drafts of the present study are gratefully acknowleged.

The computer assistance of Isabel Mendes and Ângela Pereira, and the psychological support of Bruna De Marchi are gratefully acknowledged.

Comments and help in different ways were given by Jeroen van den Bergh, Nuria Castells, Palle Haastrup, Virginie Jacquot, Teresa and Silvestro Lo Cascio, Franco Rizzari, Roman Slowinski and Giuseppe Volta. Thanks are also due to IDROSER for the collaboration provided in the case study of the Mesola wood.

Finally, I wish to thank all my multi-national friends both in Ispra and Amsterdam for having made these years so pleasant. My great hope is that our friendship will continue over the years wherever each of us will be. I have learned how easy it is to make friends of any cultural background if mutual respect exists. In my opinion, this is the most important result of these years of research!

Financial support to this research project was given by the Joint Research Centre of the European Commission, Institute for Systems Engineering and Informatics, Ispra site.

TABLE OF CONTENTS

CHAPTER 1

ENVIRONMENTAL ISSUES AND EVALUATION METHODS

1.1   Environmental Problems

The growth of world population and the rapid growth of economic activity have caused environmental stress in all socio-economic systems. There is a wide scientific consensus that problems such as greenhouse effect (and climate change), ozone depletion, acid rain, loss of biodiversity, toxic pollution and renewable and non-renewable resource depletion are clear symptoms of environmental unsustainability.

The awareness of actual and potential conflicts between economic progress in production, consumption, and technology and the environment has led to the concept of *"sustainable development"*.

The concept of sustainability has already a long history. The most widely accepted definition of *sustainable development* is the one given in the report by the World Commission on Environment and Development [1987] "Our Common Future" (known as the "Brundtland report" after the chairperson of the Commission) where sustainable development is defined as paths of human progress which meet the needs and aspirations of the present generation without compromising the ability of future generations to meet their needs[1] .

According to Daly [1991b], we can identify three of the main conflictual values of economics, allocation (efficiency), distribution (equity), and scale (sustainability). While an optimal allocation in theory could result from the individualistic marketplace, the attainment of an "optimal" scale (or at least of any scale that is not above the maximum carrying capacity) requires collective action by the community *on a regional, national or international level* according to the problems faced.

1.2   Scope and Aim of this Study

The spatial issue of environmental problems can be examined from the viewpoint of local trends causing global effects (e.g., deforestation) and global

---

[1] For an extensive discussion on the concept of sustainability see e.g.: Archibugi & Nijkamp, 1990; Barbier, 1989; van den Bergh, 1991; van den Bergh & van der Straaten, 1994; Costanza, 1991; Daly, 1991; Faucheux et al., 1994, 1995; Goodland & Ledec, 1987; Folke & Kaberger, 1991; Lovelock, 1988; Martinez-Alier, 1994; O' Connor et al., 1995; Opschoor, 1992; Opschoor & van der Straaten, 1993; Pearce et al., 1989; Pezzey, 1989; Wallace & Norton, 1992.

trends leading to local effects (e.g., acid rain) [Nijkamp et al., 1991]. The present study focuses on regional spatial dimensions of environmental management. Van den Bergh and Nijkamp [1991] argue that a meso scale of analysis is desirable for the following reasons:

- environmental decision-making can be more easily guided by a regional governmental agency;
- interactions and feedback mechanisms are more easily traceable at a meso than at a global level;
- regions have specific problems or capacities that should be dealt within their right context and level of detail.

Evaluation aims at rationalising planning and decision problems by systematically structuring all relevant aspects of policy choices (for instance, the assessment of impacts of alternative choice possibilities). Evaluation may be considered as a continuous activity which permanently takes place during the planning process. It is noteworthy that evaluation processes have often a cyclic nature. By "cyclic nature" is meant the possible adaptation of elements of the evaluation due to continuous consultations between the various parties involved in the planning process at hand. The degree of complexity of an evaluation process depends among others on the evaluation problem to be treated, the time and knowledge available and the organisational context [Munda et al., 1993a; Nijkamp et al., 1990].

It should be noted that different kinds of evaluation can be distinguished in a policy analysis, one of the most important discriminating characteristics being between monetary and non-monetary evaluation. *A monetary evaluation* is characterised by an attempt to measure all effects in monetary units, whereas a non-monetary evaluation utilises a wide variety of measurement units to asses the effects. Cost-benefit analysis and cost-effectiveness analysis are well-known examples of a monetary evaluation. Multicriteria methods belong to the family of non-monetary evaluation methods.

This book is mainly concerned with the issue of fuzzy information in multicriteria evaluation methods for environmental management problems. Given the complexity inherent in the concept of sustainable development, any method trying to operationalize this in a planning context, can be considered a kind of "second best". This is the main reason why the less ambitious concept of "environmental management" is preferred here.

Since the assumption is accepted that efficiency, equity and sustainability are the three conflictual values of economics, a mathematical procedure able to deal with these issues in an operational framework is developed.

The present study has 4 main objectives:

(1) to show the overall good theoretical and practical performance of multicriteria methods in tackling environmental problems. In particular, the consistency between multicriteria evaluation and the epistemological foundations of ecological economics is deeply investigated;

(2) to develop a new multicriteria evaluation method able to deal with crisp, stochastic and fuzzy criterion scores, in order to tackle economy-environment interactions;

(3) to develop a conflict analysis procedure aimed to be integrated with multicriteria methods, in order to operationalize equity issues;

(4) to show the empirical behaviour of such mathematical procedures by means of illustrative and real world environmental management problems.

Ecological economics explicitly recognises that economy-environment interactions are also characterised by significant institutional, political, cultural and social factors through which action is carried out. The use of several evaluation criteria is desirable. This implies that in the framework of ecological economics, the conventional economic view on evaluation has also to be changed, thus the main characteristics of cost-benefit analysis (CBA) and multicriteria evaluation (MCDA) are discussed. We show that multicriteria methods provide a flexible way of dealing with qualitative environmental effects of decisions. One should note that regarding environmental problems, CBA and MCDA can be considered as competitive methods only if all environmental consequences of decisions can be correctly transformed in monetary values; but this is very difficult. Thus we can say that, given the presence of unpriced environmental impacts, often multicriteria evaluation is the only possible approach (if also fuzzy uncertainty is considered, the use of MCDA is even more advisable).

Given the problem of the differences in the measurement levels of the variables used for economic-ecological modelling, multicriteria methods able to deal with mixed information (both qualitative and quantitative measurements) can be considered particularly useful. Traditional qualitative multicriteria approaches take into consideration the case where information on an ordinal scale is present. A problem, related to all multicriteria methods that try to take mixed information into account is the problem of equivalence of the procedures used in standardising the various evaluations of the performance of alternatives according to different criteria. Another problem related to the available information concerns the uncertainty (stochastic and/or fuzzy) contained in this information. Therefore, the combination of different levels of measurement with different types of uncertainty has to be considered as an important research issue in multicriteria evaluation.

Our overview of multicriteria evaluation methods in a fuzzy environment, shows that traditional fuzzy multicriteria methods are utility based models; most of these methods are limited to the use of only fuzzy information (often only triangular fuzzy numbers) and a key issue is how to compare fuzzy sets. Thus a new approach based on a semantic distance using areas instead of traditional intersections is presented. This new semantic distance overcomes different weak points of traditional comparison methods.

A new multicriteria method, based on some aspects of the partial comparability axiom, called NAIADE[2] (Novel Approach to Imprecise Assessment and Decision Environments) is developed. It is a discrete multicriteria method whose impact (or evaluation) matrix may include either crisp, stochastic or fuzzy measurements of the performance of an alternative a_n with respect to a judgement criterion g_m, thus it is very flexible for real-world applications. From an empirical point of view, this model is particularly suitable for economic-ecological modelling incorporating various degrees of precision of the variables taken into consideration. From a methodological point of view, two main issues are then faced:

- the problem of equivalence of the procedures used in order to standardise the various evaluations (of a mixed type) of the performance of alternatives according to different criteria;
- the problem of comparison of fuzzy numbers typical of all fuzzy multicriteria methods.

The NAIADE method presents different theoretical properties which are not shared by traditional multicriteria methods in a fuzzy environment.

Since in environmental and resource management and policy aiming at an ecologically sustainable development many conflicting issues and interests emerge, particular attention has to be given to the problem of different values and goals of different groups in society. Equity and conflicting values in multicriteria decision aid are traditionally introduced in two different ways:

(1)   *by weighting the different criteria*, but often in public decision making a single point-value solution (e.g. weights) tends to lead to deadlocks in a decision process because it imposes too rigid conditions to reach a compromise;

---

[2] Naiade is the Italian name for the Greek nymphs of rivers.

(2)   *by taking into consideration a set of ethical evaluation criteria*. A weak point of this approach is that it could lead to an excessive number of evaluation criteria. Furthermore, to identify ethical criteria may be not an easy task.

We therefore propose a third possibility i.e. the use of conflict analysis procedures to be integrated with multicriteria evaluation in order to allow policy-makers to seek for "defendable" decisions that could reduce the degree of conflict (in order to reach a certain degree of consensus) or that could have a higher degree of equity on different income groups. The *planning balance sheet method* aims at providing a broader framework for the assessment of gains and losses of a plan by constructing detailed socio-economic accounts of all project effects and by taking into account different groups in society which are affected in their well-being by the plan. A weak point of this method is that it is primarily meant to present in a systematic way a description of all the distributive impacts, but no elaboration with normative purposes is generally made. As a possible way to overcome this drawback of the planning balance sheet method we propose a fuzzy conflict resolution procedure. Starting with a matrix showing the impacts of different courses of action on each different interest/income group, a fuzzy clustering procedure indicating the groups whose interests are closer in comparison with the other ones is used. Therefore, finally a compromise solution taking into account all the three conflictual values of economics (efficiency, equity and sustainability) can in principle be identified.

The empirical relevance of the developed mathematical procedures is tested by means of a real-world environmental management problem (in the area of the Delta of the river Po in Italy). From an ecological point of view, one of the most important areas in the river Po basin is the Delta region; in this region it has been decided to establish a natural park. The Mesola wood is a part of exceptional environmental value of the Po Delta natural park. In this wood, a sharp conflict between environmental and economic aspects seems to exist. Moreover, different interest/income groups are present. From this case study the following main conclusions can be drawn:

- multicriteria evaluation can help in finding a compromise solution between conflictual ecological and economic objectives;
- the use of fuzzy sets can be a very useful tool in modelling environmental management problems characterised by deep uncertainties and approximate evaluations;
- the mathematical procedures developed in this study, can be an efficient tool to deal with efficiency aspects, equity aspects and economy-environment interactions of an environmental problem.

1.3   Multicriteria Evaluation in Environmental Management

In order to operationalize environmental management in a regional context, issues such as economic-ecological integration, multiple use, inter-regional spatial links and trade-offs, and uncertainty are of a fundamental importance [van den Bergh & Nijkamp, 1991; Munda et al., 1994].

A proper use of multicriteria analysis presupposes, the existence of an adequate environmental-economic impact system or model. Nowadays, it is increasingly taken for granted that environmental and resource problems generally have at least far reaching economic and ecological implications, often of an unpriced nature. This implies that such problems are characterised inter alia by social, psychological, physico-chemical and geological aspects. Models aiming at structuring these cross-boundary problems of an economic and environmental nature are therefore called *"economic-environmental" or "economic-ecological" models* [Braat & van Lierop, 1987; Hafkamp, 1984]. Since the complexity of this type of problems is high, there is a need for appropriate models offering a comprehensible and operational representation of a real world environmental situation. The strong quantitative tradition in economics has enabled researchers to include environmental elements fairly easily in conventional models. Nevertheless, in integrating economic and environmental models, also some methodological problems have to be faced, such as *differences in time scales* (compared to ecology, economics is mainly analysing short-term and medium term effects), *differences in spatial scales* (the spatial scale of many ecological variables is sometimes very small, whereas the scale of many economic variables is rather big) and *differences in measurement levels of the variables* (there is a clear need for methods taking into account information of a "mixed" type).

In designing models for environmental and resource policy-making the following three main types of policy objectives may be distinguished [Braat & van Lierop, 1987]:
- nature conservation objectives, e.g. "minimum exploitation of natural systems", "optimum yield";
- socio-economic objectives, e.g. "maximum production of goods and services";
- mixed objectives, e.g. "maximum sustainable use of resources and environmental services at minimum social cost".

It is clear that in policy-relevant economic-environmental evaluation models, socio-economic and nature conservation objectives are to be considered simultaneously. Consequently, multicriteria methods are in principle, an appropriate modelling tool for combined economic-environmental evaluation issues. Given the assumption of a second best world, multicriteria evaluation may be considered an appropriate tool to operationalize efficiency and sustainability criteria. This is mainly because, according to the economic-ecological integration philosophy, multicriteria evaluation allows one to tackle families of conflictual socio-economic and environmental criteria simultaneously. Given the problem of the differences in the measurement levels of the variables used for economic-ecological modelling, multicriteria methods able to deal with mixed information can be considered particularly useful.

Multiple use refers to the simultaneous use of natural resources, for different objectives, e.g. a forest which is used for outdoor recreation as well as timber production at the same time. Three broad categories of use of natural resources can be identified: *consumptive use*, *non-consumptive direct use* and *non-consumptive indirect use*. The terms consumptive and non-consumptive use are employed in an ecological sense, i.e. they refer to the resource population [Braat, 1992]. Consumptive use of a resource may of course lead to production in an economic sense, i.e. income may be derived from transforming the resource into a marketable product. This can be clarified by referring to the case of water resources management, the essential economic implication of the term use is that water is no longer suitable for subsequent desirable uses, and costs must be incurred before the water can be used again. If one type of use of a water supply creates quality deterioration partially or wholly precluding another potential use of the water, then the water has been used consumptively. An important aspect of this problem of water use compared to other economic resources is that water has a wide *quality dimension* and different qualities of water are required for different uses [Funtowicz et al., 1990].

Generally, ecosystems are used in several ways at the same time by a number of different users. This complies with the definition of multiple use. Such situations lead almost always to conflicts of interest and damage to the environment. The consequences range from sub optimal use due to unregulated access, to degradation of resource systems due to limited knowledge of the ecological processes involved. Thus, In the area of environmental and resource management and in policies aiming at an ecologically sustainable development, many conflicting issues and interests emerge. In real world situations of public decision analysis two main cases can be distinguished [Stewart, 1991]:

1) *Broad Commonalty of Goals*, i.e., differences among parties are revealed through various trade-offs which they perceive to be most in their interest.

2) *Direct Conflict of Goals*, i.e., a case where public policy involves an explicit division of resources among different sectors of the society or where attitudes have led to unreconcilable strong differences (e.g. environmentalists versus industrialists).

In the context of conflicting interests, it is also noteworthy that in environmental management there is often an interference from local, regional or national government agencies, while there is at the same time a high degree of diverging public interests and conflicts among groups in society. At an *intraregional level* many conflicting objectives may exist between different actors (consumers, firms, institutions, etc.), which can formally be represented as multiple objective problems and which have a clear impact on the spatial organisation of a certain area (e.g. industrialisation, housing construction, road infrastructure construction). At a *multiregional level* various spatial linkages exist which affect through spatial interaction and spillover effects a whole spatial system (e.g. diffusion of environmental pollution, spatial price discrimination) and which in a formal sense can be described by means of a multiple objective programming framework. At a *supraregional level* various hierarchical conflicts may emerge between regional government institutions and the central government or between regional branches and the central office of a firm, which implies again a multiple objective decision situation.

From an operational point of view, the major strength of multicriteria methods is their ability to address problems marked by various conflicting interests. Multicriteria evaluation techniques cannot solve all these conflicts, but they can help to provide more insight into the nature of these conflicts by providing systematic information into ways to arrive at political compromises in case of divergent preferences in a multi-group or committee system by making the trade-offs in a complex situation more transparent to decision-makers.

1.4   Qualitative Information in Environmental Evaluation Models

It has been argued that the presence of qualitative information in evaluation problems concerning socio-economic and physical planning is a rule, rather than an exception [Nijkamp et al., 1990]. Thus there is a clear need for methods taking into account qualitative information. In multicriteria evaluation theory, a clear distinction is made between quantitative and qualitative methods.

For the sake of simplicity, we will refer here to *qualitative information* as information measured on a nominal or ordinal scale, and to *quantitative information* as information measured on an interval or ratio scale (this last type of information is also called *crisp* information). For a concise overview of measurement scales see Appendix 1. Another problem related to the available information concerns the uncertainty contained in this information. Ideally, the information should be precise, certain, exhaustive and unequivocal. But in reality, it is often necessary to use information which does not have those characteristics so that one has to face the uncertainty of a stochastic and/or fuzzy nature present in the data [Munda et al., 1994]. If it is impossible to establish exactly the future state of the problem faced, a **stochastic uncertainty** is created; this type of uncertainty is well known it has been thoroughly studied in probability theory and statistics [e.g.Winkler & Hays, 1975].

Another type of uncertainty derives from the ambiguity of this information, since in the majority of the particularly complex problems involving men, much of the information is expressed in linguistic terms, so that it is essential to come to grips with the fuzziness that is either intrinsic or informational typical of all natural languages. Therefore, a combination of the different levels of measurement with the different types of uncertainty has to be taken into consideration[3].

Fuzzy uncertainty does not concern the occurrence of an event but the event itself, in the sense that it cannot be described unambiguously. This situation is very common in human systems. Spatial-environmental systems in particular, are complex systems characterised by subjectivity, incompleteness and imprecision (e.g., ecological processes are quite uncertain and little is known about their sensitivity to stress factors such as various types of pollution). Zadeh [1965] writes: "as the complexity of a system increases, our ability to make a precise and yet significant statement about its behaviour diminishes until a threshold is reached beyond which precision and significance (or relevance) become almost mutually exclusive characteristics" (*incompatibility principle*). Therefore, in these situations statements as "the quality of the environment is good", "the unemployment rate is low" are quite common. Fuzzy set theory is a mathematical theory for modelling situations, in which traditional modelling languages which are dichotomous in character and unambiguous in their description cannot be used. Human judgements, especially in linguistic form, appear to be plausible and natural representations of cognitive observations. We can explain this phenomenon by *cognitive distance*. A linguistic representation of an observation may require a less complicated transformation than a numerical representation, and therefore less distortion may be introduced in the former than in the latter. Fuzzy sets as formulated by Zadeh are based on the simple idea of introducing a degree of membership of an element with respect to some sets.

---

[3] A different position and an extensive criticism on the use of fuzzy sets can be found in French [1984].

The physical meaning is that a gradual instead of an abrupt transition from membership to non-membership is taken into account.

In traditional mathematics, variables are assumed to be precise, but when we are dealing with our daily language, imprecision usually prevails. Intrinsically, daily languages cannot be precisely characterised on either the syntactic or semantic level. Therefore, a word in our daily language can technically be regarded as a fuzzy set.

1.5   Structure of the Book

In light of the previous observations, this study on fuzzy information in multicriteria environmental evaluation models is organised in three main parts.

PART A

Theoretical Analysis of Cost-Benefit Analysis and Multicriteria Evaluation

Chapter 2 deals with a concise analysis of economy-environment interactions. Two main economic paradigms in approaching environmental issues i.e. conventional economics and ecological economics are compared. Since the conventional economic view on evaluation leads to the use of cost-benefit analysis, Chapter 3 is devoted to the analysis of the main features of this technique and some problems connected to its use in environmental management are emphasised. Then, Chapter 4 is concerned with the philosophy, methodology and limitations of multicriteria evaluation methods. Finally a novel systematic comparison between multicriteria evaluation and cost-benefit analysis is carried out. The argument that multicriteria evaluation is consistent with the main epistemological foundations of ecological economics is widely discussed.

PART B

Multicriteria Evaluation in a Fuzzy Environment

Chapter 5 focuses on the notion of fuzzy uncertainty, on the different ways by which it can be represented and on the open problems linked to multicriteria evaluation in a fuzzy environment. Since the comparison of fuzzy sets is one of the most important problems, Chapter 6 presents a new approach based on the use of areas instead of traditional intersections. This approach is also suitable for the problem of equivalence of the procedures used in order to standardise . different kinds of criterion scores typical of qualitative multicriteria evaluation. Thus, Chapter 7 describes a new qualitative multicriteria method, the so-called NAIADE (**N**ovel **A**pproach to **I**mprecise **A**ssessment and **D**ecision **E**nvironments) method. Its main characteristic is the possibility of dealing with crisp, fuzzy and stochastic criterion scores simultaneously.

Chapter 8 deals with the role of conflict analysis in environmental management. A fuzzy conflict resolution procedure aimed to be integrated with multicriteria evaluation is presented. As an illustrative example, the whole procedure will be applied to an environmental management problem in The Netherlands. Since the results of the NAIADE method depend on some relevant parameters, Chapter 9 focuses on the analysis of the behaviour of such a method; this is achieved by means of sensitivity analysis.

PART C

Application to a Real-World Environmental Management Problem

Chapter 10 describes the main characteristics of the river Po basin environmental management problem. Since the Po basin is a quite big area, it has been divided in a series of sub-basins for planning purposes; one of these sub-basins is the Po Delta area. In this area, the Italian government has decided to establish a natural park. Chapter 11 focuses on an environmental decision problem present in a part of the Po Delta natural park, called "boscone della Mesola".

Chapter 12 summarises the main results of this study, its limitations and possible future directions of research are indicated.

Figure 1.1 shows the structure of the book and the connections among the different chapters. According to the interests of the reader, three different possible connections among the chapters are indicated. A reader interested in both environmental management and mathematics is suggested to read the whole book. A reader with mathematical interests can only limit the reading to Chapter 1 and then from chapters 5 to 9. A reader particularly interested in environmental management is suggested to read chapters 1 - 5 and then 10 - 12.

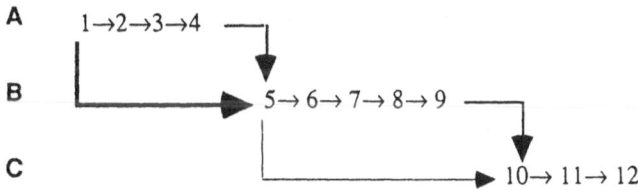

Figure 1.1   Structure of the Book

APPENDIX 1
Measurement Scales

The process of grouping individual observations into qualitative classes is measurement at its most primitive level. Sometimes this is called categorical or *nominal scaling*. The set of equivalence classes itself is called a nominal scale. The word *measurement* is usually reserved for the situation in which to each observation is assigned a number; this number reflects a magnitude of some quantitative property (how to assign this number constitutes the so-called *representation problem*).

There are at least three kinds of numerical measurement that can be distinguished: these are called ordinal scale, interval scale and ratio scale [Roberts, 1979; Vansnick, 1990; Winkler & Hays, 1975]. Imagine a set of objects O, and suppose that there is some property that all objects in the set posses, such as value, or weight, or length, or intelligence, or motivation. Furthermore, let us suppose that that each object o has a certain amount or degree of that property. In principle it is possible to assign a number $t(o)$, to any object $o \in O$, standing for the ammount that o actually "has" of that characteristic. Ideally, to measure an object o, we would like to determine this number $t(o)$ directly. However this is not always possible, therefore there is a need to find a procedure for pairing each object with another number, $m(o)$, that can be called its *numerical measurement*. The measurement procedure used constitutes a function rule $m : O \rightarrow R$, telling how to give an object o its $m(o)$ value in a systematic way. Measurement operations or procedures differ in the information that the numerical measurements themselves provide about the true magnitudes.

Let us suppose that there is a measurement procedure or rule for assigning a number $m(o)$ to each object $o \in O$, and suppose that the following statements are true for any pair of objects o_1 and $o_2 \in O$

$m(o_1) \neq m(o_2)$ only if $t(o_1) \neq t(o_2)$

$m(o_1) > m(o_2)$ only if $t(o_1) > t(o_2)$

In other words, by this rule is possible to say that if two measurements are unequal, and if one measurement is larger than another, then one magnitude exceeds another. Any measurement procedure for which both statements (1) and (2) are true is an example of *ordinal scaling*, or measurement at the ordinal level. A fundamental point in measurement theory is that of the uniqueness os scale, i.e. which are the admissible transformations of scale allowing that the truth or falsity of the statement involving numerical scales remain unchanged (*problem of meaningfulness*). In the case of an ordinal scale, it is unique up to a strictly monotone increasing transformation (with infinite degrees of liberty). Other measurement procedures associate objects $o \in O$ with a real number $m(o)$ where much stronger statements can be made about the true magnitudes from the numerical measurements. Suppose that the following statement, in addition to statements 1 and 2 is true.

$t(o) = x$ if and only if $m(o) = ax + b$, where $a \in R^+$.

That is, the numerical measurement $m(o)$ is some affine function of the true magnitude x. When the statements (1), (2) and (3) are all true, the measurement operation is called *interval scaling*, or measurement at the interval-scale level.

An interval scale is unique up to a positive affine transformation (with two degrees of freedom).

When measurement is at the interval scale level, any of the ordinary operations of arithmetic may be applied to the differences between numerical measurements, and the results can be interpreted as statements about magnitudes of the underlying property. The important part is the interpretation of a numerical result as a quantitative statement about the property shown by the objects. This is not possible for ordinal-scale numbers, but it can be done for differences between interval-scale numbers. Interval scaling is about the best we can do in most scientific work, and even this level of measurement is all too rare in social sciences. However, especially in the physical sciences, it is sometimes possible to find measurement operations making the following statement true:

$t(o) = x$ if and only if $m(o) = ax$, where $a \in R^+$.

When the measurement operation defines a function such as statements (1) through (4) are all true, then measurement is said to be at the *ratio-scale level*. For such scales, ratios of numerical measurements are unique and can be interpreted directly as ratios of magnitudes of objects.

A ratio scale is unique up to a linear transformation; in this case, the ratio between differences is unique (with only one degree of liberty).

Of course, the less the admissible transformations of a scale, the more meaningful are the statements involving that scale. From this point of view, it is better to have a ratio scale than an interval scale, and it is better to have an interval scale than an ordinal scale.

PART A

THEORETICAL ANALYSIS OF COST-BENEFIT ANALYSIS AND MULTICRITERIA EVALUATION

CHAPTER 2

FROM ENVIRONMENTAL ECONOMICS TO ECOLOGICAL ECONOMICS

2.1   Introduction

Traditional neo-classical economics analises the process of price formation by considering the economy as a *closed system*: firms sell goods and services, and then they remunerate the production factors (land, labour and capital). It is interesting to note that while classical economists such as Malthus [1798], Ricardo [1817], Mill [1857] and Marx [1867] had clear in their minds that economic activity is bounded by the environment, neo-classical economics completely forgot this important characteristic of real world economies up till to the seventies when it was started the debate on social and environmental limits to economic growth[1]. Real economy started to be seen as an *open system* that in order to function must extract resources from the environment and dispose large amounts of waste back into the environment. This is the so-called materials balance model [Ayres & Kneese, 1969; Kneese et al., 1970].

Energy analysis is very important to study the relationships between economy and environment. Energy-based valuation may appear to be a newcomer in the field of economics: some people would find its origins to the 1973 energy crisis; others identify it with Georgescu-Roegen`s "The entropy law and the economic process". Yet in actual fact, attempts to base theories of economic measurement or value on various concepts of energy have a long history behind them. Martinez-Alier [1987] shows that there is a tradition of cross-fertilisation between economics, thermodynamics and ecology due to the work of scientists as Jevons, Clausius, Podolinski, Geddes and others. However, energy is not a substitute of money in order to reach a new concept of commensurability, as it was theorised by the "energy theories of value" in the seventies and in the eighties (on this point see Faucheux & Pilet [1994]; Mirowski [1989]).

Three main economic functions of the natural system can be distinguished [Pearce & Turner, 1990]. The first function of natural environment is *to provide resource inputs to the productive system*. Economic production of any

---

[1] Here we do not enter in details regarding the so called "growth debate". The interest reader can refer to: Galbraith, 1959; Hirsch, 1977; Hueting, 1980; Meadows et al., 1972; Mishan, 1967 and 1976; Nordhaus and Tobin, 1972; Scitovsky, 1976.

commodity needs natural resources, and the transformation of natural resources, from discovery, extraction, refinement and so on, into useful raw materials and eventually into humanely produced goods and services, requires the use of industrial energy as well as the support by ecosystems driven by solar energy.

Traditionally, the economic process is considered a linear process, given the flow of natural resources, R, production P, is aimed at producing consumer goods, C, and capital goods, K. In turn, capital goods produce consumption in the future. The purpose of consumption is to create utility, U, or welfare.

Unfortunately waste (W) arises at each stage of the production process. The processing of resources creates waste in different forms; production creates waste in the form of industrial affluent and air pollution and solid waste; final consumers create waste by generating sewage, litter, and municipal refuse. While the natural systems tend to recycle their waste, economies have no such in-built tendency to recycle, thus natural environments are the ultimate repositories of waste products (2-nd function of natural environments). If the production going to create capital stock is not taken into account[2], then the *first law of thermodynamics* (stating that it is not possible to create or destroy energy and matter[3]) assures that the amount of waste in any period is equal to the amount of natural resources used up [Boulding, 1966]. According to Boulding, Earth is a closed economic system in which the economy and environment are not characterised by linear inter-linkages, but by a circular relationship; everything is an input into everything else (see Figure 2.1). The box r is recycling (it is possible to take some of the waste, W, and convert it back to resources).

As Georgescu-Roegen [1971] has pointed out, there is a basic reason for the lack of a complete recycling, this is the *second law of thermodynamics* (stating that entropy increases in any irreversible process, see Section 2.4.2.2). Given that the availability of exhaustible resources determines a finite life to the system, the circular economy will still work until the environment has the capability to absorb wastes and to convert them back into harmless or ecologically useful products. But if we dispose of waste in such a way that environment's assimilative capacity is damaged, then the function of the environment as a *waste sink* will be impaired.

---

[2] If capital formation is taken into account, in any given period a more complicated relationship between natural resources and waste exists. In fact some of the resource flows become "embodied" in capital equipment, but at the same time, capital equipment constructed in past periods will be wearing out, so it will appear as a waste flow.

[3] The first law of thermodynamics states that the total amount of energy is conserved in all processes hence in all transformations of materials. Since Einstein postulated the equivalence of matter and energy ($E=mC^2$) it follows that what is conserved is "mass-energy".

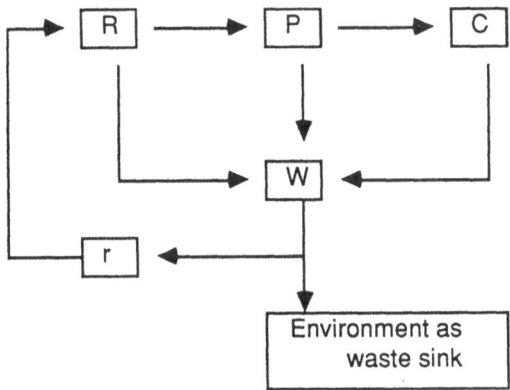

Figure 2.1   The Circular Economy
(from Pearce & Turner, 1990, p. 38)

The third function of the environment is *to supply utility directly* in the form of aesthetic and spiritual comfort. These three economic functions, resource supply, waste assimilation and aesthetic commodity, can be regarded as components of one general function of natural environments, the function of *life support* [de Groot, 1992].

The life support function of ecosystems is connected to their physical, chemical, and biological role in the overall system. Ecosystems can be divided into three categories [Odum, 1989]:

- natural environments or natural solar-powered ecosystems (open oceans, wetlands, rain forests, etc.);
- domesticated environments or man-subsided solar-powered ecosystems (agriculture lands, aqua culture, woodlands, etc.);
- fabricated environments or fuel-powered urban-industrial systems (cities, industrial areas, airports, etc.).

It is evident that fabricated environments are not self-supporting or self-maintaining. To be sustained they are dependent on the solar-powered natural and domesticated environments (life-supporting ecosystems). *Stress* caused by the disposal of wastes and pollutants, negatively affects recycling, feed-back loops and control mechanisms in the life-supporting ecosystem and thereby the production and maintenance of environmental goods and services. In the eighties, the awareness of actual and potential conflicts between economic growth and the environment has led to the concept of *"sustainable development"*.

2.2   The Concept of Sustainable Development

Traditionally, Gross National Product (GNP) has been considered as the best performance indicator for measuring national economy and welfare. But if resource depletion and degradation are factored into economic trends, what emerges is a radically different picture from that depicted by conventional methods [Daly & Cobb,1990]. In environmental terms, the GNP measure is plainly defective because [Faucheux & O`Connor, 1994]:

• no account is taken of environmental destruction or degradation;
• natural resources as such are valued at zero;
• repair and remedial expenditure such as pollution abatement measures, health care, etc., are counted as positive contribution to GNP inasmuch as they involve expenditures of economic goods and services.

Most neo-classical approaches to sustainable development define sustainability as the maintenance in the long-term of the welfare level for the population in question, where welfare is measured by *GNP per capita*. The question is typically modelled in terms of *sustainable growth* of a capital stock, treating "natural capital" as a factor of production.

Let us try to clarify some fundamental points of the concept of "sustainable development". In economics by "development" is meant "the set of changements in the economical, social, institutional and political structure needed to implement the transition from a pre-capitalistic economy based on agriculture, to an industrial capitalistic economy [Bresso, 1993|". Such a definition of development presents two main characteristics:

• the changements needed are not quantitative only (GNP growth), but qualitative too (social, institutional and political);
• there is only a possible model of development, i.e. the one of western industrialised countries. This implies that the concept of development is viewed as a process of cultural fusion toward the best knowledge, the best set of values, the best organisation and the best set of technologies.

The concept of sustainable development has wide appeal, partly because, in contrast with the "zero growth" idea of Daly [1977], it does not set economic growth and environmental preservation in sharp opposition. Rather, sustainable development carries the ideal, of a harmonisation or *simultaneous realisation* of economic growth and environmental concerns. For example, Barbier [1987, p. 103| writes that sustainable development implies: "to maximise simultaneously the biological system goals (genetic diversity, resilience, biological productivity), economic system goals (satisfaction of basic needs, enhancement of equity, increasing useful goods and services), and social system goals (cultural diversity, institutional sustainability, social justice, participation)".

This definition correctly points out that sustainable development is a *multidimensional* concept, but as multicriteria decision analysis teaches us (see Chapter 4), it is impossible to maximise different objectives at the same time.

For example, according to actual social values in western countries, to have a car per two/three persons could be considered a reasonable objective in less developed countries [Martinez-Alier, 1994a] This would imply a number of cars ten times greater than the existent one, with enormous consequences on global warming, exhaustion of petroleum, loss of agricultural land, noise, production of CO_2 and NO_x. Let us take into consideration a study by UN cited in Bresso [1993]. In 1980, the total consume of energy was of TW 10 at a world level. Without any increase in consume in less developed countries, in 2025 the whole world population would need TW 14. If the consume of the whole world population were at the level of western countries, then the consume in 2025 would be of TW 55. It is clear that while the first hypothesis is socially unsustainable (zero growth in less developed countries), the second one is environmentally unsustainable (in terms of exhaustion of natural resources and global pollution).

It is evident how difficult is to implement the idea of sustainable development. From an economic point of view, the costs and benefits of economic growth are *incommensurable* (see Chapter 3), but ecology alone (e.g. by using the concept of carrying capacity) cannot explain an important characteristic of human beings: the possibility of enormous differences in the use of materials and energy among people and territories (geographical distribution is determined historically not biologically) [Martinez-Alier, 1987, 1994a].

It is possible to synthesise the main features of sustainable development as follows. First, it is possible to find an important characteristic in the issue of *distributional equity*, both in the same generation (intra-generational equity, e.g. the North-South divide) and between different generations (inter-generational equity). Second, it is of a fundamental importance an *economic-ecological integration* (above all in terms of resource use and pollution emissions).

We could put the question, sustainable development of whom? Norgaard [1994, p.11] writes "consumers want consumption sustained, workers want jobs sustained. Capitalists and socialists have their "isms", while aristocrats and technocrats have their "cracies"". Finally, we can conclude that environmental management is essentially *conflict analysis* characterised by technical, socio-economic, environmental and political value judgments.

In the following sections we will examine how traditional neo-classical environmental economics and ecological economics differ in tackling the issue of sustainable development, in particular, the difference between *weak* and *strong* sustainability will be stressed.

2.3   Neo-Classical Environmental Economics

2.3.1   Basic Principles

Environmental economics can be considered as a particular specialisation of neo-classical economics studying two fundamental questions:
(1) the problem of environmental externalities,
(2) the correct management of natural resources (in particular, the optimal inter-generational allocation of non-renewable resources).

From an epistemological point of view, economists belonging to the Neo-Classical school take inspiration from Newton's mechanics, they tend to believe in value neutrality and objectivity and regard their arguments as "scientific". Rational decisions are connected with the existence of optimal solutions based on calculations in *monetary* or other *unidimensional* terms. Central premises of this economic paradigm are [Klaassen & Opschoor, 1991]:
*   the maximisation premise,
*   the weighting premise,
*   the fixed content premise.

The *maximisation premise* on behaviour states that "economic men" (individuals and groups) try to maximise their objective function (especially welfare for individuals and profit for enterprises) and individual welfare judgments are the ultimate criterion. According to this assumption, politicians are assumed to maximise their votes and bureaucrats their financial budget or power position in other respects. Egoistic motives are assumed to dominate. The economic value of marketable commodities, unpriced environmental goods and services, or sympathy for future generations, is determined according to the amount of personal utility yielded. The preferences of individuals are revealed by the decisions they make and *efficiency and consistency* of decisions reflect rational behaviour. The preference structure is assumed to hold only the *preference* and the *indifference* relations (in Chapter 4 the consequences of these assumptions will be investigated).

The weighting premise on evaluation states that all relevant changes as a consequence of economic decisions can be expressed in a welfare-related, *one-dimensional entity*, so that costs and benefits of all alternatives can be reduced to neat (ordinal) balance figures that can be ranked (assumption of complete commensurability). It has to be noted that to put a precise monetary value to an environmental externality implies to solve very important problems such as uncertainty connected to the environmental impact, correct time horizon and correct discount rate. A concise discussion of such monetary valuation methods will be presented in Chapter 3.

The *fixed content premise* states that a range of parameters are assumed to be static or given, including: institutional arrangements (in especially the economic system), preferences and needs, the state of technology and the state and functioning of the environment. The consequences of this premise are particularly evident in the attitude that economists belonging to this school have towards economic growth.

Such economists have a quite optimistic view of technological progress and economic growth. They generally recognise that even if the production technologies of an economy can potentially yield increases in output commensurate with increases in inputs, overall output will be constrained by limited supplies of resources (growth theory with exhaustible resources). But these limits can be overcome by *technological progress*: if the rate of technological progress is high enough to offset the decline in the per capita quantity of natural resource services available, output per worker can rise indefinitely. A stronger statement is the following: *even in the absence of any technological progress exhaustible resources do not pose a fundamental problem if reproducible man-made capital is sufficiently* substitutable *for natural resources* [Dasgupta & Heal, 1979, Hartwick, 1977, 1978; Solow, 1974, Stiglitz, 1979].

From a macroeconomic point of view, generally production is written as $Y=f(K, L, R)$, i.e., output is a function of man-made capital, labour and resource stocks. But, since the production function is almost always of a multiplicative form, such as the Cobb-Douglas, R can easily approach zero with Y constant if only K or L are increased in a compensatory fashion [Daly, 1991b].

The Cobb-Douglas production function is:

$$Y=K^{\alpha}R^{\beta}L^{\gamma} \qquad \text{with } \alpha, \beta, \gamma > 0 \text{ and } \alpha+\beta+\gamma=1 \qquad (2.1)$$

where Y is output, K capital, R resources, and L labour.

The Cobb-Douglas production function is used because it presents two important properties:

(1) the natural resources are essential (if R=0, then Y=0);

(2) the average product of natural capital does not have an upper bound (i.e. the limit for R→ 0 of Y/R is equal to infinity).

These properties are not shared by CES functions since in the case of an elasticity of substitution greater that 1, no input is essential, and in the case of an elasticity smaller than 1, the average product of each of the inputs has an upper bound (for a formal proof see Arrow et al. [1961]).

The crucial question is whether or not $\alpha > \beta$ (where these two parameters represent the elasticities of output with respect to reproducible capital and the exhaustible resources); roughly speaking, if $\alpha > \beta$ fixed capital is sufficiently important in production to allow for possibility of a permanently maintainable output level despite the declining availability of the natural resources. Dasgupta and Heal [1979, p. 205] arrive at the conclusion that "one would imagine α to be about four times β". When the premise of technological progress is introduced, the possibilities for sustained increases in output improve even more. The problem is that a backstop technology[4] is simply *assumed to exist*, thus the technological progress is assumed to provide always the right solutions!

This concept of substitution of more productive man-made capital for natural capital can be criticised from many sides.

(1)   The Cobb-Douglas production function does not allow an analysis of the possible limits to substitution among inputs because *a constant degree of substitution between inputs is simply assumed* [Victor, 1991].

(2)   If capital depreciates by a constant proportion, the exhaustible resources are essential, since consumption should eventually fall to zero (assuming no technical change). This limitation is usually recognised, but considered "highly unrealistic" [Dasgupta & Heal, 1979].

(3)   Man-made capital is not independent of natural capital; since resources are required to manufacture capital goods the success of any attempt to substitute capital for resources will be limited by the extent to which the increase in capital requires an input of resources. "The idea of substitution might be rescued if we can demonstrate that the extra productivity in K_M (man-made capital) outweighs the extra natural resources that get used up in the production of K_M. At this stage all we can say is that this is not obvious [Pearce & Turner, 1990, p. 49]".

---

[4] A backstop technology is an almost indefinitely renewable resource which eventually takes over when exhaustible resources have gone (e.g., energy from fusion reactors).

(4)   A limit to the substitutability between man-made capital and natural capital is that natural capital has the feature of multifunctionality (all the life support functions), such a feature is not shared by man-made capital [Pearce & Turner, 1990].

(5)   Production functions of a Cobb-Douglas type confuse "flow" variables with "fund" variables. For example, the input of labour referred to as L represents flow of services provided by the work force. Similarly, R is a measure of the flow of services from natural resources. K, however, is a measure of the stock of capital and not a flow of services. Hence, L, R, and K are not directly comparable [Georgescu-Roegen, 1979].

The so called *weak sustainability concept* [Pearce & Atkinson, 1992, 1993] states that an economy can be considered sustainable if it saves more than the combined depreciation of natural and man-made capital. "We can pass on less environment so long as we offset this loss by increasing the stock of roads and machinery, or other man-made (physical) capital. Alternatively, we can have fewer roads and factories so long as we compensate by having more wetlands or mixed woodlands or more education [Turner et al., 1994, p. 56]". Weak sustainability is based on a very strong assumption, *perfect substitutability* between the different forms of capital, then all the criticism presented above also applies in this case.

Under weak sustainability conditions, sustainability is equivalent to leaving future generations with a total stock of capital not smaller than the one enjoyed by the present generation. Cabeza [1995] notes that the concept of weak sustainability is nothing but a by-product of growth theory with exhaustible resources when:

(1) the definition of inter-generational equity is restricted to a non-declining level of consumption per capita;

(2) the relationship environment-economy is restricted to the introduction of an aggregate input called natural capital into the production function.

Indeed, weak sustainability is simply a different statement of the so called Hartwick-Solow rule [Hartwick, 1977, 1978; Solow, 1974, 1986], stating that in order to have a stream of constant level of consumption per capita to infinity, society should invest all the current returns from the utilisation of the flows from the stock of exhaustible resources.

Finally, we can synthesise the position of standard economic theory towards sustainable development in the following propositions:

production is the result of the combination of capital, labour, and natural resources with pollution as an externality. Growth of consumption can be sustained even if production and consumption deplete a natural resource faster than it regenerates if:

- the resource can continuously be substituted for capital, or
- if there is exogenous resource-saving technical progress.

If production/consumption results in pollution, the level of production that can be sustained will be lower due to negative impacts of pollution on productivity, resource regeneration and the fact that part of production capacity has to be sacrificed for investments in pollution control equipment [Klaassen & Opschoor, 1991].

2.3.2 A Different View: Maintaining the Natural Capital Stock

Pearce and Turner [1990] although they are inside the framework of conventional economics[5], have a different position in approaching environmental problems. They devote their attention to the desirability and meaning of maintaining the natural capital stock as a condition for sustainable development. Maintaining the natural capital stock is considered *desirable* for the following reasons:

(1)   Complete substitution between natural capital and man-made capital is impossible;

(2)   There is no guarantee that new technology is necessarily less polluting. Furthermore, it is uncertain that technological progress will continue forever, or at least for a very long time. "There may indeed be backstop technologies that will free us from natural resources, but they cannot be brought into existence simply by assuming that they are there [Pearce & Turner, 1990, p. 50]".

(3)   The role which natural environments play in supporting and sustaining economic systems is covered by scientific uncertainty. Since *uncertainty* exists on the way in which environments function, either internally or in terms of their interactions with the economy, a trade-off of the benefits of substituting man-made capital for natural capital is not a serious one. Moreover, most environmental decisions are characterised by irreversibility, if a mistake is made, it is not possible to correct it afterwards (it is quite difficult to create again a

---

[5] "In this textbook we show how we can use the main body of economic thought to derive important propositions about the linkages between the economy and the environment. Rather than looking for some "different economics", we are seeking to expand the horizons of economic thought [Pearce & Turner, 1990, p. 30]".

tropical forest). Thus the presence of *uncertainty and irreversibility* together should make human beings more circumspect about giving up natural capital.

(4)   The poorest countries in the world rely on natural resources far more directly than advanced economies. The sustainability of these societies depends on the maintenance of the stocks of these natural resources[6].

(5)   Another reason for maintaining the resource stock is to ensure broad access to it by different generations. Any irreversibility now means the removal of an option for future generations, they cannot secure access to the resource if it has been made extinct.

(6)   If the assumption that animals have rights is accepted, one of those rights must be the existence in order to exercise other rights. When natural capital is destroyed, also the habitat that wild animals require for their existence is destroyed.

But what does a constant natural capital stock *mean?* Pearce and Turner [1990, p. 53] give four possibilities:

- the physical quantity of natural resource stocks should remain unchanged;
- the total value of the natural resource stocks should remain constant in real terms (standard economic approach);
- the unit value of the services of the natural resources, as measured by the prices of natural resources, should remain constant in real terms;
- the value of the resource which flows from the natural resource stock should remain constant in real terms. Where resource flow is the product of price and quantity used, thus it is possible to allow quantity to decline but the price to rise, keeping value constant.

Pearce and Turner recognise some of the shortcomings of each of these definitions of a constant stock of natural capital, other weak points have been indicated by Victor [1991]. Measurements of natural capital stock made exclusively in physical terms are problematic because of the difficulty in adding up different physical quantities expressed in different units. For this reason the second interpretation is offered. By valuing each resource stock in money terms, the total value of natural capital can be measured. One obvious problem here is that many natural resources (e.g., air, water, wilderness) do not have observable prices. Thus one would need to find implicit or shadow prices in some way. Even those prices that do exist may not be useful; they may be affected by market

---

[6] Sustainable development implies a concern for the future, and this concern is necessarily reduced by discounting. Everyone discounts the future to some extent; generally speaking, the poor people are forced to discount at a higher rates than the well off. The need for immediate survival outweighs any consideration of the future. There is some poverty level below which sustainability becomes an unaffordable luxury. Thus poverty gives birth to an over exploitation of natural resources [Clark, 1991].

imperfections and taxes, and they may exclude externalities involved with the production and use of the resource. "There are additional problems in using market prices to value the aggregate stock of natural capital. Resource prices or net prices reflect conditions at the margin and to use these to value entire stocks can give perverse results. For example, it is possible for the real price or net price of a resource to rise over time at the same rate as (or faster than) the rate of decrease in the physical stock of the resource..... This possibility is of more than theoretical interest. If price or net price rises as resource quantity is declining, the value of resource stocks as an indicator of sustainability can give precisely the wrong policy signal to government. As long as the value of the stock remains constant or rises, the government, through this indicator, will not perceive a problem even though the flow of resource is becoming increasingly valuable (as measured by price) and the physical stock is declining [Victor, 1991, p. 204]". Pearce and Turner's third and fourth interpretations of a constant stock of natural capital also utilise market prices and so similar criticism made in relation to keep the value of the capital stock constant apply. These problems are unlikely to be overcome easily. Indeed the problem of measuring capital has been one of the fundamental sources of criticism of conventional economics levelled by the Post-Keynesian school[7]. The so-called *"Cambridge Controversy"* [Harcourt, 1972] deals with the problem of measurability of capital. Capital here is referred to man-made capital, but the results can easily be extended to natural capital. The *quantity of capital* depends on its value (price), *its value* depends on the rate of interest (the maximum price that a buyer will pay for a capital good is the present value derived from the increase in output over time that is made possible by the acquisition of the capital good, such a present value depends on the interest rate), and the *rate of interest* (price of capital which is determined in the capital market) depends on the *quantity of capital!*

The difficulties involved in finding theoretically sound, robust measures of the stock of natural capital may be even greater than those identified by the Post-Keynesians for manufactured capital.

Although the idea of a constant natural capital stock is quite important and desirable (maintaining the natural capital is an important prerequisite for sustainability), one should admit that the above considerations demonstrate that the development of relevant indicators of sustainable development connected to

---

[7] Keynes and in general Post-Keynesians, have paid few attention to the problem of economy-environment interactions. Some interesting relations between some aspects of Post-Keynesian theory and Bioeconomics are highlighted by Gowdy [1991]. Although some common points can be found (methodological framework, emphasis on production rather than exchange, interpretation of social rate of discount), an important potential area of conflict between the two schools is their attitude toward economic growth.

this idea is quite difficult, this mainly because it is based on the assumption of complete commensurability.

2.4   Ecological Economics

The linkages between ecosystems and economic systems are the focus of ecological economics. A good definition of what is meant by Ecological Economics is the following. "Increasing awareness that our global ecological life support system is endangered, is forcing us to realise that decisions made on the basis of local, narrow, short-term criteria can produce disastrous results globally and in the long run. We are also beginning to realise that traditional economic and ecological models and concepts fall short in their ability to deal with global ecological problems. *Ecological economics* is a new trandisciplinary field of study that addresses the relationships between ecosystems and economic systems in the broadest sense..... Ecological economics (EE) differs from both conventional economics and conventional ecology in terms of breadth of its perception of the problem, and the importance it attaches to *environment-economy interactions* [Costanza et al., 1991, pp. 2-3]".

A simplified scheme of the possible scientific approaches to environment-economy interactions can be found in Figure 2.2. The left half concerns those approaches using several evaluation criteria for analysing the interactions between ecological and economic systems, and the right half those using a common denominator for this evaluation, such as money or energy. Ecological economics explicitly refuses the complete commensurability paradigm and recognises the existence of *incommensurability* between economic and environmental aspects [O`Neil, 1993], thus a new scientific paradigm is needed.

2.4.1   Epistemological Foundations

2.4.1.1   Ecological Economics as a Post-Normal Science

In any science a paradigm or pre-analytic vision exists; research has to start somewhere, thus something is given by a pre-analytic cognitive act. Everybody starts his own research from the work of his predecessors. According to Kuhn [1962], scientists normally are just ordinary people (so neither the impeccable truth-gathers of the positivist tradition, nor the heroic conjecturalists of Popper) concerned only in solving research puzzles within an *unquestioned* framework of concepts and methods.

Global environmental issues present new tasks for science; scientists now tackle problems introduced through policy issues where typically, *facts are uncertain, values in dispute, stakes high, and decisions urgent* [Funtowicz & Ravetz, 1991, 1994]. Thus Funtowicz and Ravetz have developed a new epistemological framework called "post-normal science", where it is possible to make use of two crucial aspects of science in the policy domain: *uncertainty* and *value conflict*. The name "post-normal" indicates that the puzzle-solving exercises of normal science, in the Kuhnian sense, which were so successfully extended from the laboratory of core science to the conquest of nature through applied science are no longer appropriate for the solution of environmental problems.

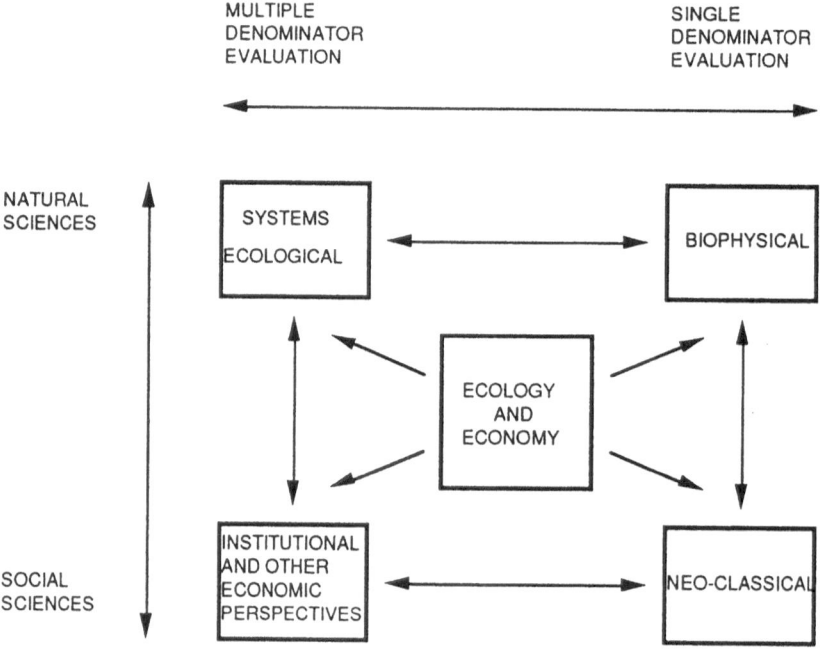

Figure 2.2   A Simplified Conceptual Model of Ecological and Economic
Perspectives and Approaches to Environmental Issues
(from Folke & Kaberger, 1991b, p. 275)

Neo-classical economics has traditionally been able to maintain its credibility by relegating uncertainties in knowledge and complexities in ethics firmly to the sidelines. But, uncertainties in input information produces irreducible uncertainty in conclusions; the relevant question of quality is the degree to which the recommended policy choices are robust against those underlying uncertainties. As a post-normal science, ecological economics recognises the presence, importance and legitimacy of different value-commitments for the appropriate management of uncertainty. It does not claim ethical neutrality, nor an indifference to the policy consequences of its arguments.

When science became used in policy, it was discovered that lay-persons (e.g. judges, journalists, scientists from another field, or just citizens) could master enough of the methodology to become effective participants in the dialogue. A basic principle of post-normal science is that these new participants are indispensable. This extension of the peer community is essential for maintaining the quality of the process of resolution of complex systems. Thus the appropriate management of *quality* is enriched to include this multiplicity of participants and perspectives. The criteria of quality in this new context will, as in traditional science, presuppose ethical principles. But in this case, *the principles will be explicit and will become part of the dialogue.*

One should note that the view that concerning environmental issues, *conflicts* between interests and interested parties are the normal state of affairs is also shared by institutional economics [Myrdal, 1973, 1978; Söderbaum, 1992].

According to Funtowicz and Ravetz [1994], the traditional analytical approach, implicitly or explicitly reducing all goods to commodities can be recognised as one perspective among several, legitimate as a point of view and as a reflection of real power structures, but not the whole story. To choose any particular operational definition for value involves making a decision about what is important and real; other definitions will reflect the commitments of other stakeholders. How much is a songbird worth? To answer this question represents a new problem of valuation, one where measurements cannot pretend to be independent of methodology and ethics. "The issue is not whether it is only the marketplace that can determine value, for economists have long debated other means of valuation; our concern is with the assumption that in any dialogue, all valuations or "numeraires" should be reducible to a single one-dimension standard [Funtowicz and Ravetz, 1994a, p. 198]". In Chapter 3 the issue of commensurability and its criticism will be extensively discussed.

In ecological economics, instead of the "maximisation premise" and the "weighting premise", a *multidimensional* approach to economic and policy-making is emphasised. In this book, the introduction of multicriteria decision aid techniques in the framework of ecological economics is considered desirable. Since such techniques are based on a "constructive" rationality (see Chapter 4) and allow one to take into account conflictual, multidimensional, incommensurable and uncertain effects of decisions, they can be considered perfectly consistent with the methodological foundations of ecological economics.

2.4.1.2   The Coevolutionary Paradigm

There is a constant and active *interaction* of the organisms with their environment; organisms are not simply the results but they are also the causes of their own environments, this is the main thesis of the *coevolutionary paradigm* [Norgaard, 1988, 1994]. Economic development can be viewed as a process of adaptation to a changing environment while itself being a source of environmental change. However, coevolution does not imply change in a particular direction (i.e., progress).

In biology, coevolution refers to the pattern of evolutionary change of two closely interacting species where the fitness of the genetic traits within each species is largely governed by the dominant genetic traits of the other. In real world societies, "people survive to a large extent as members of groups. Group success depends on culture: the system of values, beliefs, artifacts, and art forms which sustain social organisation and rationalise action. Values and beliefs which fit the ecosystem survive and multiply; less fit ones eventually disappear. And thus cultural traits are selected much like genetic traits. At the same time, cultural values and beliefs influence how people interact with their ecosystem and apply selective pressure on species. Not only have people and their environment coevolved, but social systems and environmental systems have coevolved [Norgaard, 1994, p.41]".

Agriculture began between five and ten thousand years ago when there were approximately five million people in the world. Population doubled eight times, increasing to about 1.6 billion people by the middle of the nineteenth century. These eight doublings were only possible through an increase in the effectiveness with which people interacted with their environment through changes in knowledge, technology and social organisation. According to Norgaard, the increase in material well-being and in the rate of population during the past century and half can be understood as a process of coevolution. With industrialisation, social systems evolved to facilitate development through the exploitation of coal and petroleum[8] . Social systems no longer coevolved to interact more effectively with environmental systems. "Hydrocarbons freed societies from immediate environmental constraints but not from ultimate environmental constraints - the limits of the hydrocarbons themselves and of the

---

[8] Martinez-Alier [1987] shows that the increase in productivity of modern agriculture depends on the underestimation of energetic inputs of fossil fuels, the low value given to the contamination caused by pesticides and the loss of biodiversity.

atmosphere and oceans to absorb carbon dioxide and other greenhouse gases associated with fossil fuel economies [Norgaard, 1994, p.44]".

From the coevolutionary paradigm the following lessons can be learned:

(1) A priori, different models of coevolution are possible, then no unique optimal development path exists. The spatial dimension is a key feature of sustainable development;

(2) the respect of cultural diversity is of a fundamental importance. In environmental management local knowledge and expertise (being the result of a long coevolutionary process) sometimes are more useful than experts` opinions;

(3) coevolving systems have parts and relations which change in unforeseeable ways. At any point in time, they can be described like an ecosystem, but over time they are as unpredictable as the evolution of life itself.

It has to be noted that the principles of the coevolutionary paradigm and post-normal science reinforce each other. They share the issues of value conflicts, democratisation of science and uncertainty. Post normal science emphasises the importance of incommensurability and decision making processes; coevolution underlines the importance of economy-environment interactions.

2.4.2   Economy-Environment Interactions

2.4.2.1   The Issue of Scale

In ecological economics, the "fixed content premise" is replaced by one of circular interdependence incorporating the major processes in the environment and taking into account essential biophysical laws.

Systemic approaches to environmental issues consider the relationships between three systems: the economic system, the human system and the natural system [Passet, 1979]. The *economic system* includes the economic activities of man, such as production, exchange and consumption. Given the scarcity phenomenon, such a system is efficiency oriented. The *human system* comprises all activities of human beings on our planet. It includes the spheres of biological human elements, of inspiration, of aesthetics, and of morality which constitute the frame of human life. Since it is clear that the economic system does not constitute the entire human system, one may assume that the economic system is a subsystem of the human system. Finally, the *natural system* includes both the human system and the economic system [Nijkamp & Bithas, 1992].

From the ecological economic perspective, the expansion of the economic subsystem is limited by the size of the overall finite global ecosystem, by its dependence on the life support sustained by intricate ecological connections which are more easily disrupted as the scale of the economic subsystem grows relative to the overall system. Since the human expansion, with the associated exploitation and disposal of waste and pollutants, not only affects the natural environment as such, but also the level and composition of environmentally produced goods and services required to sustain society, the economic subsystem will be limited by the impacts of its own actions on the environment [Folke, 1991]. A central issue then is: does any *"optimal" scale* exist for the economy? This point has especially been tackled by Daly.

The term scale is shorthand for "the physical scale or size of the human presence in the ecosystem, as measured by population times per capita resource use [Daly, 1991b, p. 35]". The standard economics point of view about economic growth seems quite optimistic. But as an economy grows, it increases in scale. Scale has a maximum limit defined either by the *regenerative or absorptive capacity of the ecosystem*, therefore "until the surface of the earth begins to grow at a rate equal to the rate of interest [Daly, 1991b, p. 40]", one should not take this optimistic attitude too seriously. Thus, a sharp difference exists between the philosophy of "growth" referring only to quantitative scale of the physical dimensions of the economy, and the philosophy of "development" referring only to qualitative improvement. Then we can speak of a *steady-state economy* [Daly, 1977, 1991a] as one which develops without growing.

2.4.2.2 The Entropy Law and the Economic Process

Since the meaning of the entropy law for the economic process is a quite discussed subject, here we will follow closely Georgescu-Roegen terminology.

Classical thermodynamics deals with energy but only with energy in bulk. No thermodynamic concept makes any sense if applied to a microscopic element. An electron has no heat, no temperature, no pressure, and no entropy. The entropy concept can be defined as follows: "in an isolated thermodynamic system the available energy continuously and irrevocably degrades into an equal quantity of unavailable energy, so that the total energy remains constant while the unavailable energy keeps increasingly up to a maximum [Georgescu-Roegen, 1993, p. 187]". Where, available energy is the one that humans could use for their purposes while unavailable energy is energy that humans cannot use in any way; an *isolated system* is the one that can exchange neither energy nor matter with its environment. The entropy law can be applied neither to a *closed system* that can exchange only energy with the environment nor to an *open system* that can exchange both energy and matter with its outside.

As matter exists, like energy, in two states: available and non-available, and since matter-energy enters the economic process in a state of low entropy and comes out of it in a state of high entropy, Roegen states his discussed fourth law as follows: "in a closed system (as the Earth practically is) mechanical work cannot proceed at a constant rate forever [Georgescu-Roegen, 1993, p. 198]", or more simply, unavailable matter cannot be completely recycled. A consequence of this law is that a program based on the substitution of terrestrial energy by solar energy as Daly`steady state, cannot work. Interesting discussions of the meaning and consequences of the fourth law can be found in Mayumi [1991, 1992, 1993].

Regarding technological progress, Georgescu-Roegen is quite pessimistic. He defines "Promethean techniques" as the ones that allow to obtain a surplus of *accessible* energy (to get more accessible energy than that used in the operation). A new technology requires a new Promethean technique, not just one already familiar alternative. The Promethean technique that saved the wood crisis is the steam engine, but nowadays neither the controlled fusion nor the direct harmessed solar energy have the characteristics of a Promethean technique.

Figure 2.3 illustrates the ecological economics conception of the economic system as a part of the overall ecosystem.

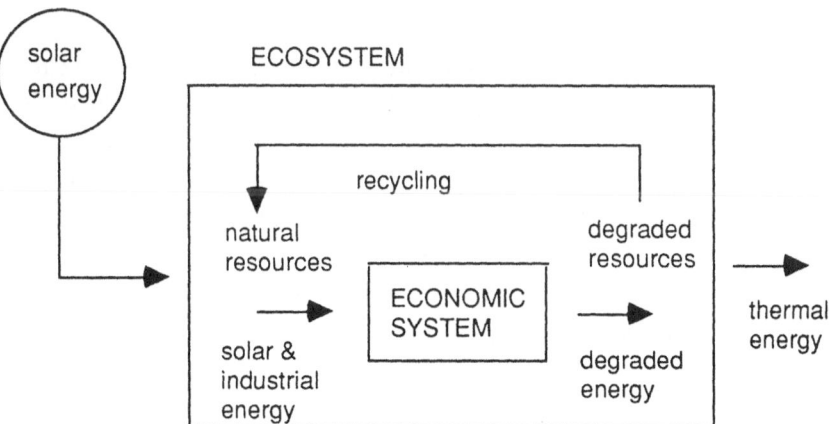

Figure 2.3 Economy as a Part of the Ecosystem

2.4.3 What is Ecological Economics?

First of all it is an economics that accepts that the scale of the economy has a maximum limit defined either by the regenerative or absorptive capacity of the ecosystem; thus the concept of *"strong sustainability "* is used. Such a definition is based on the assumption that certain sorts of natural capital are deemed critical, and not readily substitutable by man-made capital [Barbier & Markandya, 1990]. In particular, the characterisation of sustainability in terms of the "strong" criterion of non-negative change over time in stocks of specified natural capital provides a strong justification for development of non-monetary indicators of ecological sustainability based on *direct physical measurement* of important stocks and flows [Faucheux & O`Connor, 1994]. What is then an ecological economy?

An ecological economy is an economy that keeps the exploitation of renewable resources (fish, trees, water) within the limits of their rates of renewal; it is an economy that uses non-renewable resources (e.g. petroleum) in line with achievement of a transition (through substitution, technological change, etc.) to a base of renewable resources. It is an economy that keeps the production of waste and contamination within the assimilative capacity of the environment. An ecological economy is an economy that preserves biodiversity in all its expressions. It is an economy that recognises that both ecological and economical rationality are not sufficient to lead to correct decisions, thus environmental decisions must be taken by using a democratic scientific-political decision process.

In conclusion, one should note that traditionally, the theoretical focus of economics is on prices and the issue is to internalise external environmental costs to arrive at prices that reflect full social marginal opportunity costs. In this way, in theory the problem of *efficient allocation* can be solved. Under ideal conditions the market can find an optimal allocation in the sense of Pareto. Another problem is the *just distribution*; the market's criterion for distributing income is to provide an incentive for efficient allocation, not to attain justice. These two values can conflict, and the market does not automatically resolve this conflict. Generally there is agreement that it is better to let prices serve efficiency, and to serve equity with income distribution policy. But the market cannot find an optimal scale any more than it can find an optimal distribution. The latter requires the addition of *ethical criteria*, the former requires the addition of *ecological criteria* [Daly, 1991b].

According to Kuhn, scientific progress alternates between "normal" and "revolutionary" phases, in which (respectively) scientists make piecemeal advances, or choose between rival grand systems. But how one can choose in a revolutionary situation? Is there any criterion by which "genuine progress" of science can be accounted for? Of course, Kuhn offers no methods or criteria for helping scientists decide in a revolutionary situation. One should note that conventional economics can arrive at elegant and clear-cut conclusions, ecological economics however shows that assumptions on technical progress and substitution need to be revised in the light of biophysical and ecological laws. Fruitful results could be expected from a co-operation between the two schools of thought and thus the two paradigms should not be considered as mutually exclusive.

2.5 Conclusions

The following consequences may be drawn:

(1) Natural life-supporting ecosystems are negatively affected by the disposal of wastes from the economic system. If the economy-environment interactions are taken into account, immediately a broad question about the capability of the natural environments to sustain the economy arises.

(2) Substitution of more productive man-made capital for natural capital is not an acceptable answer to environmental problems.

(3) The idea of maintaining the natural capital stock is important and desirable; unfortunately it is very difficult to operationalize. Its main problem is connected to the possibility of valuating environmental goods in money terms. If such a valuation were possible, sustainability could be operationalized by taking into account a portfolio of projects. In this case, individual projects may use environmental resources as long as this is compensated elsewhere in the set of projects.

(4) Environmental problems are very complex and characterised by scientific uncertainty. Any method trying to operationalize the concept of sustainable development is necessarily a second best approach.

(5) In economic theory three main conflictual values can be identified: allocation, distribution and scale. In an operational framework, this means that an exhaustive analysis has to take into consideration efficiency criteria, ethical criteria and ecological criteria, thus a multidimensional paradigm is needed.

(6) Ecological economics explicitly recognises that economy-environment interactions are also characterised by significant institutional, political, cultural and social factors through which action is carried out. The use of several evaluation criteria is desirable. This implies that in the framework of ecological economics, the maximisation and the weighting premises of neo-classical

economics have also to be changed. Since multicriteria techniques are based on a "constructive" rationality and allow one to take into account conflictual, multidimensional, incommensurable and uncertain effects of decisions, they can be considered perfectly consistent with the methodological foundations of ecological economics.

CHAPTER 3

COST-BENEFIT ANALYSIS

3.1   Introduction

In Chapter 2 it has been shown that according to Neo-Classical economic theory, decision processes are based on two premises. The *maximisation premise* on behaviour states that "economic men" (individuals and groups) try to maximise their objective function (especially welfare for individuals and profit for enterprises) and individual welfare judgements are the ultimate criterion. *The weighting premise* on evaluation states that all relevant changes as a consequence of economic decisions can be expressed in a welfare-related, one-dimensional entity, so that costs and benefits of all alternatives can be reduced to neat (ordinal) balance figures that can be ranked. These two premises imply the use of monetary evaluation methods. *A monetary evaluation* is characterised by an attempt to measure all effects in monetary units, whereas a non-monetary evaluation utilises a wide variety of measurement units to asses the effects. Cost-benefit analysis and cost-effectiveness analysis are well-known examples of a monetary evaluation.

The history of plan and project evaluation before World War II showed first a strong tendency towards a financial trade-off analysis. Later on much attention was focused on cost-effectiveness principles. After World War II, cost-benefit analysis gained increasing popularity in public policy evaluation, by using willingness to pay notions, consumer surplus principles and shadow prices. Social cost-benefit analysis (CBA)[1] can be regarded as an effective kind of applied welfare economics. It consists of the following main steps [Bojö et al., 1990; Dasgupta & Pearce, 1972; Hanley & Spash, 1993; Mishan, 1971a; Pearce, 1971; Pearce & Nash, 1989]:
- identification of costs and benefits
- quantification and evaluation of costs and benefits in terms of a common monetary unit
- choice of a social rate of discount
- choice of a time horizon
- construction of a one-dimensional indicator bringing together all the benefits and costs (many authors suggest the use of the net present value).

---

[1] In this study we do not consider private projects, thus by CBA is always meant Social Cost-Benefit Analysis.

The social returns are composed of all gains and losses of all members of society whose well-being will be affected by the plan if implemented. These gains and losses are measured by the preferences of the individuals who are affected. The use of such conventional optimisation models has been criticised from many sides. The optimising approach is based on the assumption that different objectives can be expressed in a common denominator by means of trade-offs, so that the loss in one objective can be evaluated against the gain in another. This idea of *compensatory changes* underlies both the classical economic utility theory and the traditional cost-benefit analysis. The determination of a common denominator is, however, fraught with difficulties. Furthermore, in the past decades, the degraded state of the natural environment has become another key issue in evaluation because of the externalities involved and it is increasingly taken for granted that environmental and resource problems generally have at least far reaching economic and ecological aspects, which cannot always be encapsulated by a market system. The limits inherent in conventional evaluation methodologies and the necessity of analysing conflicts between policy objectives have led to a need for more appropriate analytical tools for strategic evaluation, such as multicriteria evaluation. However, we do not conjecture that conventional monetary evaluation tools are to be discarded. It is noteworthy that the debate on conventional evaluation analysis and multidimensional analysis tends to regard these two methodologies increasingly as complementary analytical tools (which may reinforce each other) than as competitive methods. Seen from this perspective, a blend of monetary and non-monetary evaluation tools may be a fruitful direction in evaluation research. CBA remains useful as one of the possible inputs to decision making, as long as policy makers bear its limitations in mind.

In the next sections we will give a more detailed discussion of some characteristics and limitations of cost-benefit analysis.

3.2   The Philosophical Foundations of Cost-Benefit Analysis

The rationality behind cost-benefit analysis assumes that any individual makes rational decisions only if he weighs up the advantages and disadvantages of a particular action, so that some kind of best decision can always be made. Of course, the essence of cost-benefit analysis is that it is not confined to decisions that affect one individual; it relates to social decisions. Then, does the characteristic of rationality remain if we extend it to the social context? The basic argument underlying CBA is that this rationality does remain. That is, if we leave a whole group of individuals to carry out their own personal cost-benefit analyses in respect to a given policy, then we can simply aggregate the results to secure a social evaluation.

It is the emphasis on this social view that generates many of the philosophical problems of CBA. First, what set of individuals constitutes society? More important, only the individuals of the present society are counted. On this argument, CBA could be undemocratic if it is judged on behalf of future generations.

Second, cost-benefit analysis tends to equate the social view with what society wants. The notion of individual preference that is relevant to CBA is the preference that is recorded in the market place (or which would be recorded if there were a market), and not the preference recorded by a simple vote. The result of a cost-benefit analysis, therefore, will not necessarily coincide with the results of a simple majority voting procedure. There is a problem then of making CBA compatible with democratic decision making [Pearce, 1971]. This kind of "economic voting" is preferred to classical political voting procedures for five main reasons [Pearce & Nash, 1989]:

(1) political systems other than in very well defined referenda, involve voting not for issues so much as for individuals to represent the constituent's view. Market or economic voting is far more true to the voters' intentions: by definition if the voter does not want it, he does not buy it;

(2) even if referenda were desirable, they cannot be held everyday on every economic decisions that has to be made;

(3) political voting is very much an affair influenced by personalities and, often, purely irrational factors. By economic voting individual items can be separated out and voted on separately;

(4) failure to vote in an election simply results in the outcome being determined by those who decide to vote. But in the market a failure to vote is a vote against the project: it will show up in a benefit figure lower than it would otherwise;

(5) the use of money values permits some expression of the intensity of preference in the vote: it enables the individual to say how deeply he wants or does not want the project or goods in question.

However, it is generally recognised that such differences are not of a clear-cut type. For example, the absence of the politicisation of issues dealt with in the marketplace may not always be a good thing, e.g. the introduction of a new drug cannot be determined simply by individuals' demand. Market votes are also open to influence from the "hidden persuaders" of advertising, and so on.

A basic idea of conventional economics is that prices of goods and services are signals reflecting consumer desires. The free interplay of market forces should therefore allocate resources to goods and goods to people in such a way as to secure the maximum welfare of society, where welfare is equated with the satisfaction of wants. No modern economist seriously believes that any real world economy operates in such a way as to maximise welfare. Rather, the idea that there exist some configurations of prices which will achieve this optimum is used as a yardstick against which to measure the degree of imperfection in an economy and hence the extent to which policies should be directed towards correcting those deviations.

Upon which principles the measurement of social costs and benefits is to be based? Before the rise of ordinalism, the principle seemed obvious: measure the total utility (disutility) caused to all members of society.

The *consumer surplus* is a frequently used concept in a cost-benefit analysis in order to judge whether the project in question provides a net contribution to raising the level of aggregate consumption. Consumer's surplus is the difference between what an individual has to pay for a good (the market price) and what an individual would be willing to pay for each unit of the good rather than to go without it. From the utility maximisation problem it follows that a Marshallian demand function can be viewed as a locus of points representing the individual's willingness to pay values for each unit of the market good. Hence consumer's surplus is defined as the area under the demand function and above the price line. For environmental goods and services which have no market price, the consumer's surplus is defined as the area under the demand function (and above the zero price line). A necessary condition for an effective calculation of the aggregate consumer surplus is knowledge of the demand curves for the elements of the project at hand. Unfortunately, frequently little information is available about these demand curves.

The use of consumer's surplus can be criticised from many sides. First the assumption of a linear demand curve needs to be accepted. Second the assumption of constant marginal utility of income is introduced [Samuelson, 1942; Patinkin, 1963]. Thus, if no account is taken of the differing marginal utilities of income across persons, willingness to pay clearly depends upon the *ability to pay*. Projects which benefit higher income groups will therefore tend to appear more attractive than projects which benefits lower income groups.

In recent years, much attention has been devoted to the more general question of how to evaluate consumer surpluses in the multiple price change case (path-dependency issue). The basic problem is that the sum of the changes in consumer surpluses in general depends on the order in which prices are changed. However, conditions for path independency have been established, and these conditions turn out to be closely related to the aformentioned constancy of the marginal utility of income [McKenzie & Pearce, 1982; Morey, 1984]. Third, Little [1950] has concluded that consumer's surplus is no more than a "totally useless theoretically toy" on the ground that the demand curve is only partial and fails to take account of the effect of the investment on the prices of all other goods, i.e. there will be changes in surplus elsewhere which are not accounted for by the analysis of the project in question.

With the rise of ordinalism, the problem has become even more complex. If utility is regarded solely as an ordinal concept, how - even in principle - can the disutility imposed on different members of society be aggregated? The solution which has been most commonly adopted is the so called *"compensation principle"* usually associated with the names of Hicks [1939] and Kaldor [1939]. By this, the social cost of a given output is defined as the sum of money which is just adequate when paid as compensation to restore to their previous level of utility all who lose as a result of production of the output in question. In other words, the Kaldor-Hicks principle declares a social state y "socially preferable" to an existing social state x if those who gain from the move to y can compensate those who lose and still have some gains left over. Such a situation is consistent with a Pareto improvement since we have x indifferent to y for the losers (once they are compensated) and y preferred to x for the gainers (if they can over-compensate). It is just this principle which underlies cost-benefit analysis. If the monetary value of benefits exceeds the monetary value of costs, then the gainers can hypothetically compensate the losers and still have some gains left over. The excess of gains over required compensation is equal to the net benefits of the project.

Since the compensation principle was formulated, it has been attacked from several sides. Amongst the most important contributions to the debate are those of Scitowsky [1941] who first noted the possibility that the undertaking of a project without the payment of compensation may redistribute income in such a way that an ex post application of the compensation test yields a different answer from an ex ante one, and Little [1950] who stressed the value content of the approach and the need to take distributional factors into account.

The Kaldor-Hicks test requires only that gainers be able to compensate losers, *it does not require actual payment to be made*. Scitovsky demonstrated that in absence of compensation, it is possible for circumstances to exist such that once the change has come about, a move back to the status quo can also be judged socially desirable. In essence what happens is that the change is desirable when valued at the new set of prices that emerge from the new distribution of income resulting from the policy change. Since, in general, no mechanism exists for the transfer of funds from beneficiaries to losers, the Scitovsky paradox rises considerable doubts upon the usefulness of the Kaldor-Hicks formula, and as a consequence upon this aspect of the welfare foundations of cost-benefit evaluation [Johansson, 1987].

Many decisions will lead to widespread price changes, resulting in some consumers paying more for goods they purchase, and others less. Scitovsky [1954] has termed such effects *pecuniary externalities*. Price changes themselves redistribute income; for every consumer who pays more, a producer receives more, and vice versa. Therefore, if we are adopting the compensation principle, such changes are to be ignored. Once again, it is necessary to stress the lack of concern for *distributional questions* embodied in this way of measuring social costs.

Generally, it is said that cost-benefit analysis focuses on *efficiency* criteria; *equity* problems are ignored. But, any policy decision affects the welfare positions of individuals, regions or groups in different ways; consequently, the public support for a certain policy decision will very much depend on the *distribution effects* of such a decision. For an extensive discussion on the non-separability between allocative efficiency and distribution see Martinez-Alier and O'Connor [1995].

Given that society is unlikely to be indifferent between various distribution of income, some ways of integrating the distributional aspects into the analysis have to be found. Three approaches can be distinguished:
- to ignore the issue without further comment. This attitude is rarely defended, but often practised [Bojö et al., 1990];
- to introduce distributional weights explicitly;
- to use the planning balance sheet method.

Some revisions of cost-benefit analysis try to include distribution values directly in the analysis by using different weights for different social groups [Helmers, 1979]. The main limit of this approach is that it is not clear how to derive such weights and who should attach these ones. It has been argued that the economist - qua economist - has no right to attach these social utilities of income of individuals. Thus, the decision-maker has to assign them. However, to asses such weights is a delicate problem (in Chapter 4 it will be shown that this is an important problem also in multicriteria evaluation). In any case, if weights are used, it has to be recognised that no completely objective analysis is possible, and therefore no optimal solution exists. Finally, it has to be noted that failures to use any weighting system implies making the value judgements that the existing distribution of income is optimal. If, and only if one is happy with such a value judgement, it is reasonably possible to use unweighted market valuations to measure costs and benefits. *Therefore, there is no escape from value judgements.*

The *planning balance sheet method* (also called "community impact analysis") can be considered an extension of traditional cost-benefit analysis [Lichfield, 1964, 1988, 1993]. This method aims at providing a broader framework for the assessment of gains and losses of a plan by constructing detailed socio-economic accounts of all project effects and by taking into account different groups in society which are affected in their well-being by the plan. In this way both the efficiency and equity aspects can be considered.

This approach requires the construction of a matrix showing the various alternatives and the impacts of these alternatives on different income groups. It is not necessary that all gains and losses are represented in monetary units; even an ordinal scale can be used.

A problem inherent in this method is that the priorities of the various groups are hardly traded off against each other, so that a serious problem of weighting schemes still remains. This method is primarily meant to present in a systematic way a description of all the distributive impacts, but no elaboration with normative purposes is generally made.

An obvious advantage of the balance sheet method is that multiple groups are taken into account. Finally, it should be noted that since the balance sheet method is not a pure monetary evaluation method, it can be considered as something in between traditional cost-benefit analysis and multicriteria evaluation.

3.3 Valuation Approaches of Costs and Benefits

If all costs and benefits can be measured in money terms, then it is possible to adopt a simple rule:
Maximise (Benefits - Costs).
The above formulation can be written more rigorously as:

$$\max \sum_{t}^{T} \frac{B_t - C_t}{(1 + r)^t} \qquad (3.1)$$

where B_t and C_t are benefits and costs in time period t, r is the discount rate and T is the time horizon. Obviously, if B_t and C_t are in different units, no such decision rule can be used. Indeed, when it proves practically impossible to secure monetary estimates of a project's output (whether cost or benefit), the analysis is reduced to cost-effectiveness analysis.

In order to be consistent with the objective of maximising social welfare, it is necessary that the prices attached to the physical benefits and costs reflect society's valuations of the final goods and resource involved. Two questions immediately arise:

(1) If markets do exist, to what extent will observed market prices reflect social valuations?

(2) If markets do not exist (as it happens for most environmental goods and services), how are surrogate prices to be derived which, in turn, reflect social valuations?

The main valuation techniques use [Bojö et al., 1990]:

(1)   conventional markets,

(2)   implicit markets,

(3)   artificial markets.

3.3.1 Conventional Markets

The fact that conventional markets are used does not necessarily mean that market prices are adopted without alterations. When significant distortions are present, appropriate shadow prices have to be estimated.

In classical welfare economics, prices resulting from a competitive equilibrium can be considered to be a measure of social opportunity costs. Deviations from the neo-classical model originate from the so-called *"market failures"*. Market distortions such as monopoly, taxes, price regulations and disequilibria often play an important role in the economy. As a result, prices may be bad indicators of the real scarcities and pertaining social evaluations in the economy. Some set of prices, called *shadow or accounting prices*, which reflect the true social opportunity cost of using resources in a given project need to be computed. In general, we would expect the marginal cost of a final good to indicate society's valuation of that good, since the marginal cost reflects consumers' willingness to use resources in that situation. As a first approximation, shadow prices are assumed to reflect marginal costs. Of course, shadow prices should reflect marginal social costs rather than marginal private costs. However, the use of marginal cost pricing in the public sector with prices elsewhere diverging from marginal costs involves the "second best problem". The essential argument is that setting prices equal to marginal cost in one sector only may actually move the economy away from a Pareto optimum. In other words, given that a "perfect" is not achievable (prices equal to marginal costs everywhere), marginal cost pricing in the public sector will not guarantee a "second best" (i.e. the best available position given that a "perfect" cannot be secured) [Lipsey & Lancaster, 1956].

Clearly, if market prices are to be corrected so that they reflect marginal costs, there is a *practical* problem of estimating marginal costs and a *conceptual* problem of justifying the procedure in the face of the second best theorem. Furthermore, marginal private cost will still not fulfil the role of a proper shadow price if private and social cost diverge. An important cause of divergence is the presence of an important category of market failures contributing to environmental degradation, viz. *externalities* [Ayres & Kneese, 1990; Mishan, 1971b; Nijkamp, 1980]. In order to deal with the problem of consequences that are not priced at all on a market, neo-classical economists use the concept of externalities[2]. Pollution can then be considered as an external diseconomy. The necessity of operationalizing the externalities concept in environmental management has led to the following typology of theoretical responses to externalities:
(1)   optimisation,
(2)   compensation,
(3)   internalisation.

As noted by Verhoef [1993], consensus on the exact definitions and interpretation of these concepts seems to be lacking in literature. However, the following definitions are offered by such an author [Verhoef, 1993, p. 6]:
• an externality is *optimised* when its level is consistent with Pareto efficiency according to the Kaldor-Hicks criterion;
• an externality is *compensated* when a (financial) transaction takes place between the supplier and the receptor of the effect, which compensate for the welfare effects due to the externality;
• an externality is *internalised* if a market for the effect comes into being.

Clearly, optimisation is an efficiency related concept, whereas compensation is an equity related concept. However, one has to note that the general problems of the relation between efficiency and equity in CBA, discussed earlier apply also in this case. For environmental problems the Kaldor-Hicks principle can be

---

[2] A useful definition of externality has been suggested by Baumol and Oates [1975, p. 17]: "An externality is present whenever some individual's (say A's) utility or production relationships include real (that is, non monetary) variables, whose values are chosen by others (persons, corporations, governments) without particular attention to the effects on A's welfare".

formalised in a simple way as follows. Economic theory states that the utility derived from consume can be captured in a so called utility function, e.g., U=U(M, E), where M represents monetary income and E is environmental quality. Given such a utility function, indifference curves are defined as the locus of points representing combinations of money and environmental quality that yield the *same level of utility*. Indifference curves result from several neoclassical axioms of choice, the most important being completeness (no incomparability relation exists), transitivity (if **a** is indifferent to **b** and **b** is indifferent to **c**, then **a** and **c** are indifferent), convexity (indifference curves are strictly convex to the origin) and non-satiation (more will always be better). These axioms are introduced in order to formulate any decision problem in terms of utility maximisation and to guarantee a unique solution to this maximisation problem. In Chapter 4 a criticism of such axioms will be illustrated.

This model implies that it is always possible to find an amount of money in terms of willingness to pay for environmental quality improvements or of willingness to accept for environmental quality deteriorations that keeps utility constant. This implies two main consequences:

(1)   The optimisation and compensation models can be regarded as crucial tools in conventional economics, because only in this way one may assign an amount of money to environmental decay. However, it has to be noted that such models do not aim at achieving a better environmental quality, but only at incorporating the environmental impacts in the traditional price and market system. After internalisation, market forces will take over, and thus no room for political intervention exists. It has to be noted that since the objective is to keep utility constant, complete substitution between environmental quality and economic growth is always allowed, then a *weak sustainability* philosophy is implied.

(2)   The compensation model presents strong distributive impacts; the monetary value of a negative externality depends on social institutions and distributional conflicts. If the people damaged are poor or of future generation, the cost of internalisation will be lower ("the poor sell cheap") [Martinez-Alier, 1994b, 1994c].

Martinez-Alier and O'Connor [1995] have introduced the concept of *ecological distribution*, referring to the social, spatial, and temporal asymmetries or inequalities in the use by humans of environmental resources and services. Thus, the territorial asymmetries between SO_2 emissions and the burdens of acid rain is an example of *spatial ecological distribution*; the intergenerational inequalities between the enjoyment of nuclear energy and the burdens of radioactive waste is an example of *temporal ecological distribution*. In the USA, "environmental racism" means locating polluting industries or toxic waste disposal sites in areas of Black or Hispanic or Indian population. This is an example of *social ecological distribution*.

Finally, one should note that since externalities are characterised by the absence of markets, there will also be an absence of observable prices with which the cost-benefit analysis can work. Many external effect problems therefore reduce the issue of valuing "intangibles". This leads to the use of implicit and artificial markets.

3.3.2 Implicit and Artificial Markets

The basic idea behind *implicit markets* is that there are links between the consumption of ordinary goods sold on markets and the consumption of non-marketed goods, including environmental values. Thus, changes in environmental quality are also reflected in prices of ordinary goods, such as land and houses. But sometimes it is not possible to make inferences from actual behaviour; thus one may have to measure consumer preferences in hypothetical situations or by creating *artificial markets*. This approach is often called contingent valuation.

In many applications of CBA to environmental issues, it is necessary to place monetary values on non-market goods such as clean air, clean water and wilderness areas. Several methodologies have been developed to cope with such estimation requirements, the principal ones being contingent valuation, the travel cost method, hedonic pricing, and the shadow project approach. Among these only *contingent valuation* is universally applicable. The aim of contingent valuation is to elicit valuations (or "bids") which are close to those that would be revealed if an actual market existed. Respondents say that they would be willing to pay or willing to accept if a market existed for the good in question. In order to determine the value of environmental goods and services, economists try to identify how much people would be willing to pay (willingness to pay (WTP)) for these goods in artificial markets. Alternatively, the respondents could be asked to express their willingness to accept (WTA) compensation.

The respondents must be familiar with the good in question and with the hypothetical means of payment (payment vehicle). The quality of results in this method depends on how well informed people are, moreover, the problem with these techniques is that respondents may answer "strategically". For example, if they think their response may increase the probability of implementing a project they desire, they may state a value higher than their true value (free rider problem). In order to avoid free rider behaviour people should really pay the amount of money they indicate; unfortunately in this case, WTP depends upon the *ability to pay*, thus projects which benefit higher income groups would generally considered to be the best. Furthermore, society as a whole, may have values that deviate from aggregated individual values. Society has a much longer life expectancy than individuals, thus the value society attaches to natural resources and the environment is likely to deviate from individual values, since the simple summation of individual preferences may imply the extinction of species and ecosystems. This implies that environmental policy *cannot be merely based upon the aggregation of individual values, and estimation of willingness to pay at any particular point of time* [Klaassen & Opschoor, 1991]. Thus, willingness to pay measures can be criticised from both the intratemporal and intertemporal points of view.

One interesting aspect of contingent valuation studies is the difference found empirically between WTA and WTP measures. How are the differences to be explained? There are various options, the main ones are the following [Pearce & Turner, 1990; Hanley, 1992]:

(1)   people value gains and losses "asymmetrically", an individual is not indifferent to a given gain and an equivalent loss. This behaviour is explained by psychologists by means of *prospect theory*. In prospect theory individuals' values relate to gains and losses in comparison to some reference points. This contrasts with the economic assumption that individuals maximise utility. What matters is the point from which the gains and losses are measured.

(2)   WTP bids are constrained by income, whereas WTA bids are not (my WTP to save my life is bounded by my income and ability to borrow; my WTA is probably infinite).

(3)   Empirical problems in collecting WTA estimates in contingent valuation exist. Economists have argued that WTA bid formats suffers excessively from hypothetical market bias, respondents find it difficult to accept the notion of compensation for environmental damages, or for improvements that do not occur. There is much evidence of a significant "protest bid" in WTA formats. Sagoff [1988] argues that what protest bidders are signalling in refusing to state minimum compensation sums for a loss of environmental quality is that no amount of money can compensate them for such losses. "If some environmental assets are indeed priceless, on WTA measures, then this rules out CBA as a methodology for deciding on their level of provision as the Kaldor-Hicks criterion rests on the possibility of compensation for losers" [Hanley, 1992, p. 37].

The divergence between WTA and WTP measures of value is a serious problem. This phenomenon can lead to projects passing the Kaldor-Hicks test when payment measures are elicited for welfare losses, but failing it when compensation measures are obtained.

The *travel cost method* is most frequently used to estimate consumers' surplus for recreation sites, by using travel costs as a proxy for price, and then deriving the relationship between visit rates and the cost of visiting [Anderson & Bishop, 1986]. Thus travel cost models are based on an extension of the theory of consumer demand in which special attention is paid to the value of time. This methodology presents the following problems:

- the choice of the dependent variable;
- the treatment of respondents who visit more than one site on the day out;
- the treatment of holiday-makers' travel costs;
- determination of the value of time;
- a very large amount of information is often required and so simplifying assumptions will be necessary in many cases;
- the applicability of the methodology is very limited.

The *hedonic price approach* works by finding a relationship between goods and the value of the attributes of the goods. Thus, the site value can be considered the capitalisation of location attributes including residential tax liabilities and environmental qualities. Given that different locations have varied environmental attributes, such variations will result in differences in property values. With the use of appropriate statistical techniques, the hedonic approach attempts to:

- identify how much of a property differential is due to a particular environmental difference between properties, and
- infer how much people are willing to pay for an improvement in the environmental quality that they face and what the social value of improvement is.

The identification of a property price effect due to a difference in pollution levels is usually done by means of a multiple regression technique. Of course, the differences in residential property values can arise from any source, and hence such studies usually involve a number of property variables, a number of neighbourhood variables, a number of accessibility variables and finally the environmental variables of interest.

Such a methodology presents the following problems:

- If any variable that is relevant is excluded from the analysis, then the estimated effects on property value of the included variables may be biased. On the other hand, if one includes as many variable as possible, typically many of the variables of interest are themselves very closely correlated, and thus multi-collinearity will occur;
- the assumption of an efficiently operating housing market is needed.

An extensive discussion on the main properties of these valuation methods can be found in Hoevenagel [1994].

Another adaptation of traditional cost-benefit analysis is the *shadow project approach*. The idea is that the costs of deterioration of a natural area or of a historical building can be assessed from the costs of creating an equivalent project elsewhere (a so-called "shadow project"). The shadow project need not necessarily be actually implemented; it has only significance as an indirect step to gauge the costs of intangible losses of the original project. It is clear that a basic problem of the shadow concept is the definition of an equivalent project. Certain projects are unique as the result of a long historical, cultural or ecological development, so that the *time dimension* plays a crucial and sometimes prohibitive role in the definition of a shadow project. In addition, the *spatial dimension* must not be neglected, because the value of a certain project is co-determined by its accessibility. If the shadow project has a different accessibility, the compensating costs must be corrected for travel time differences. One should be aware of the fact that a shadow project has only a concrete meaning if its site is known. The creation of a shadow project at a different place affects in turn the land use at that place; thus here again, a second shadow project would have to be defined in order to calculate the intangible losses due to the shadow project. In this way, a whole chain of shadow projects might be defined, which probably would lead to an indeterminate solution.

The concept of a shadow project is of fundamental importance to answer the question whether CBA is consistent with a goal of sustainable development. If the Pearce and Turner definition of sustainable development is accepted, the answer is yes. This is providing that the government receives sufficient shadow projects to offset environmental damages, so that across a portfolio of public investments, net environmental damage is zero. However, besides the aggregation problems inherent in this definition of sustainability extensively discussed in Chapter 2, there are problems here, both in measuring environmental impacts and in designing shadow projects which fully compensate. Moreover, allowing for shadow projects may even increase the level of environmental degradation, since natural resources fulfil many functions, and the future consequences of shadow projects may be many and uncertain [Munro & Hanley, 1991; Nijkamp & van Delft, 1977].

Finally, one has to note that a frequent ethical criticism of CBA is that natural resources, human life and health are not economic assets and thus they cannot be valued in economic terms. We do not maintain that a human life has infinite value; for example, a reduction in road accidents can be secured at some cost, but society is unlikely to devote the whole of the national income to this end. Logically, any intangible has a value, but *in practice* the derivation of this value may be impossible; intangible and incommensurable effects are very hard to incorporate in cost-benefit analysis. Therefore, the conclusion is justified that any attempt to transform a priori heterogeneous and unpriced impacts into a single dimension runs the risk of failure.

3.4   The Social Rate of Discount

When a multi-period monetary analysis is carried out, it is necessary that the valuations of costs and benefits are computed in real values (constant prices) instead of nominal values (current prices). Thus, there is a need for discounting. It has to be noted that the problem is not simply the issue of relative prices (i.e., inflation), but the fact that people find it natural that the value of one dollar today is greater than the value of one dollar in 10 years. There are two underlying reasons for the existence of positive interest rates [Bojö et al., 1990; Nijkamp & Rouwendal, 1987]:

(1) people discount the future because they simply prefer their benefits now to later; they are impatient (time preference). This means that a postponement of current consumption toward a future period is only acceptable if its resulting welfare loss is compensated for by means of a rate of discount.

(2) The second source of interest rates is the productivity of capital. A dollar could be invested now and could therefore be worth more in real terms in 10 years (there is an opportunity cost in terms of return on capital foregone).

If these two ideas are translated to the level of society, two different main approaches exist.

The first is called the social time preference rate (STPR) approach. This assumes the rate to be mainly a political parameter set on the basis of (a) per capita income growth perspective, (b) the rate at which utility of increases in marginal income diminishes and sometimes, (c) an assumption of pure rate of time preference among consumers [Mueller, 1974; Pearce, 1983; Sen,1961, 1967; Solow, 1974].

The second is the social opportunity cost of capital (SOC) approach which looks for empirical evidence of profits on alternative investments opportunities [Arrow, 1966; Baumol, 1968; Hirschleifer,1966; Stiglitz, 1982].

Unfortunately, over the years economists have found no objective discount rate to be used in a cost-benefit analysis; on the contrary, the choice of a discount rate is one of the most disputed subjects of economic theory. However, one thing is sure, all discount rates are greater than zero. Some people regard this as immoral, simply because it does appear to be inconsistent with the ideas of conservation and sustainability (the higher the discount rate, the faster the resources are likely to be depleted, thus discounting contains an in built bias against future generations). However, to manipulate the discount rate for environmental reasons may generate many problems (for example, a low discount rate makes more projects produce a positive net present value; this may cause an increase in pollution and natural resource degradation because of increased investment activity). Here, we do not wish to go into details in this subject; however, it is necessary to stress the fact that the choice of a time horizon and a discount rate can influence very much the results of a cost-benefit analysis. In any case, it has to be noted that this problem of discounting is not only connected with CBA, but it is present in any analysis that is intended to take future consequences of alternative courses of action into account.

3.5 The Construction of a One-Dimensional Performance Indicator

When the analysis is of a monocriterion type (i.e. construction of a single criterion), from a mathematical point of view, it is possible to find an optimal solution. However, it has to be noted that in order to speak of an optimum, the problem must be formulated in such a way that the following three properties are satisfied [Roy & Bouyssou, 1991]:

(1) the alternatives are all mutually exclusive;
(2) the set of alternatives is well-defined and fixed;
(3) the alternatives can be ranked in an incontestable way from the worst to the best (of course indifference relations are allowed).

The investment evaluation criterion generally used in order to compare inter temporal costs and benefits is the *net present value* (NPV). In financial economics, a distinction is made between *absolute and relative investment criteria* [Brealey & Myers, 1985; Matarazzo, 1981]. The net present value, since it is an homogeneous linear function of the cash flows, is an indicator of the financial convenience of an investment in absolute terms. That is, if all the (monetized) costs and benefits of a given project I are multiplied by a constant k, the new project kI has a value of the net present value k times bigger than the one of I, i.e.,

$$NPV(kI) = k\ NPV(I) \qquad \forall\ k \in R^+ \tag{3.2}$$

It has been shown [Hirshleifer, 1958] that the NPV represents the right evaluation criterion only if the projects considered are mutually independent. Other investment criteria such as the benefit-cost ratio or the internal rate of return are homogeneous functions of degree zero of the cash-flows and therefore, the financial convenience of a project is independent from the dimensions of such a project. These brief considerations make clear that the assumptions underlying each investment criterion are different and as a consequence also the results they generate are different. This can be easily shown graphically [Matarazzo, 1981, pp. 191-205]. If for simplicity we assume positive cash flows equal to $\bar{Y}$ and negative cash flows equal to $\bar{X}$ we have:

$$NPV = \bar{Y} - \bar{X} \qquad while \qquad benefit\text{-}cost\ ratio = \frac{\bar{Y}}{\bar{X}}$$

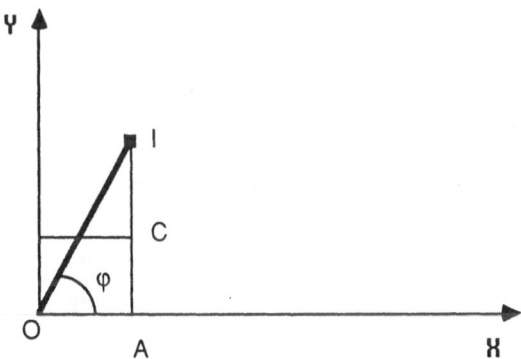

Figure 3.1   Graphical Representation of
NPV and Benefit-Cost Ratio of a Project I

In Figure 3.1, it is clear that given the project I, benefit-cost ratio $= \dfrac{\bar{Y}}{\bar{X}} = \tan \varphi$ while $NPV = \bar{Y} - \bar{X}$ is given by the straight-line AC.

Thus in Figure 3.2, one can see that according to the benefit-cost ratio, since $\varphi_1 > \varphi_2 > \varphi_3$, the ranking of alternatives is $I_1 > I_2 > I_3$, while according to the net present value it is $I_2 > I_1 > I_3$.

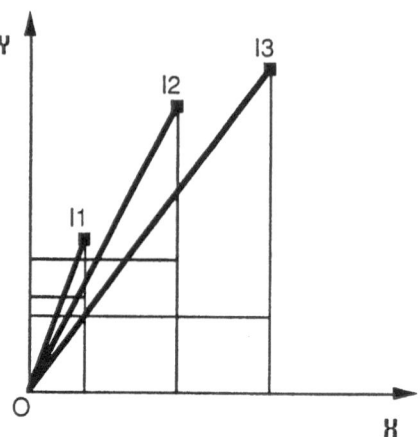

Figure 3.2   Ranking of Different Projects
According to Different Criteria

The decision problem faced by a decision-maker may be a simple "accept-reject" decision or it may also be one of choosing between competing alternatives. In the former case both NPV and benefit-cost ratio give the same results. In the latter case, if only one action has to be chosen, then the choice should be made on the basis of the highest NPV; another decision context arises where a number of projects can be chosen but the budget available is limited, the rule then is to rank the projects according to the benefit-cost ratio [Turner et al., 1994].

Finally, one has to note that both NPV and benefit-cost ratio can be considered to be additive utility functions, thus the properties indicated in Section 4.7.1.1 should always hold.

CHAPTER 4

MULTIPLE CRITERIA EVALUATION METHODS

4.1   Introduction

Environmental management is essentially conflict analysis characterised by technical, socio-economic, environmental and political value judgements. Therefore, in an environmental planning process it is very difficult to arrive at straightforward and unambiguous solutions. This implies that such a multi-related planning process will always be characterised by the search for acceptable compromise solutions, an activity which requires an adequate evaluation methodology. Multiple criteria evaluation techniques aim at providing such a set of tools. Multicriteria methods provide a flexible way of dealing with qualitative multidimensional environmental effects of decisions. However, this does not mean that multicriteria evaluation is a panacea which can be used in all circumstances without difficulties; it has its own problems.

In the past, the field of operational research used to formalise decision-making problems as follows:

(1) a well defined set A of feasible alternatives;

(2) a criterion defined on A reflecting the preferences of the decision-maker precisely;

(3) a well formulated mathematical problem:

e.g. to find a* in A such as

$$g(a^*) \geq g(a) \quad \forall \, a \in A \tag{4.1}$$

where g is a real valued function defined in A.

The choice of a* represents the optimum decision, because a* is better than all other alternatives in A. The mathematical techniques for finding the solution of a problem formulated in this way, are well known.

During the last two decades, it has increasingly been understood that welfare is a multidimensional variable which includes, inter alia, average income, growth, environmental quality, distribution equity, supply of public facilities, accessibility, etc. This implies that a systematic evaluation of public plans or projects has to be based on the distinction and measurement of a broad set of criteria. These criteria can be different in nature: private economic (investment costs, rate of return, etc.), socio-economic (employment, income distribution, access to facilities, etc.), environmental (pollution, deterioration of natural areas, noise, etc.), energy (use of energy, technological innovation, risk, etc.), physical planning (congestion, population density, accessibility, etc.) and so forth [Nijkamp et al., 1990]. Multicriteria evaluation can be considered as the evolution of the preceding theory: in fact it enables the use of several decision criteria (i. e. not only just a function g, but more than one).

A great number of multicriteria methods has been developed and applied for different policy purposes in different contexts [Bana e Costa, 1990a; Bell et al., 1988; Bogetoft & Pruzan, 1991; Fandel et al., 1983; Janssen, 1992; Keeney & Raiffa, 1976; Nijkamp et al., 1990; Rietveld, 1980; Roy, 1985; Roy & Bouyssou, 1993; Schärlig, 1985; Steuer, 1986; Vincke, 1989; Yu, 1985; Zeleny, 1982; Zionts, 1982].

In general, a multicriteria model presents the following aspects:
(1) There is no solution optimising all the criteria at the same time and therefore the decision-maker has to find *compromise solutions.*
(2) The relations of preference and indifference are not enough in this approach, because when an action is better than another one for some criteria, it is usually worse for others, so that many pairs of actions remain incomparable with respect to dominance relation.

The main advantage of these models is that they make it possible to consider a large number of data, relations and objectives (often in conflict) which are generally present in a specific real-world decision problem, so that the decision problem at hand can be studied from multiple angles.

The main disadvantage of a multicriteria model is that an action **a** may be better than an action **b** according to one criterion and worse according to another. Thus when different conflicting evaluation criteria are taken into consideration, a multicriteria problem is mathematically ill-defined. The consequence is that a complete axiomatization of multicriteria decision theory is quite difficult. As Vincke [1985] has observed, in these cases the following attitudes are unproductive:
"1) leave the decision-maker entire liberty for the decision,
2) introduce consciously or not restrictive hypotheses, so that the problem can be solved by a classical method. The methods used in multicriteria analysis lie between these two extremes: they are based on models constructed partly from necessarily restrictive mathematical hypotheses, and from information gathered from the decision-maker".

4.2 Philosophical Foundations of Multiple Criteria Decision Aid

Two opposite positions in tackling decision problems may be distinguished [Munda, 1993]:

Decisionism in practice maintains that decisions are blind actions, inspired by the subconscious and by the instincts, so that the act of reasoning over a decision is meaningless.

On the contrary, *rationalism* assumes that in any decision problem an optimal precise solution always exists and that it is possible to find it by reasoning over the problem. Thus (using Socrates' words) ignorance is the only cause of foolish or evil acts.

Multicriteria models have the aim, with the aid above all of the mathematical instrument, of leading to concrete decisions. Of course multicriteria models are not pure *axiomatic systems* but beyond the applicational aspect, multicriteria evaluation represents a specific mathematical theory and, as such, should respect the general principles proposed by the formalistic school and therefore indicate clearly the axioms on which the development of the model should depend. Now from this observation, it is possible to deduce that, like in non-Euclidean geometries, the development of the different hypotheses and therefore the formulation of the various models multiply at great speed.

Multicriteria models are *factual* in the sense that the hypotheses on which they are based must be close to reality, but an approach of an exclusively descriptive type does not appear suitable for the purposes that decision analysis should have. This type of approach is common to those who assume an essentially platonistic attitude, believing in an "objective world" which exists on its own behalf and in optimal solutions that the researcher has only to discover. An approach of this type may be useful in particular cases, such as some problems of technical optimisation, but in general it is not considered desirable. Those assumptions which hypothesise the ability of the decision-maker to express in clear terms his utility function, or his absolutely consistent preferences of a transitive type, do not in fact appear to be very realistic, as a consequence an evident antinomy is formed with a hypothesis of descriptivity.

The *prescriptive approach* [1] does no more than propose a series of rules which the decision-maker should respect if he wishes to reach a specific objective. Simon [1983] proposes a behavioural model according to which human rationality is "bounded" and rational decision making only requires the application of a set of personal values to solve specific problems a person faces, in a way

---

[1] Here we do not distinguish between normative and prescriptive models [Bell et al., 1988].

that is "satisfactory" for that person. Thus the concept of "decision process" has an essential importance. "In general it is impossible to say that a decision is a good one or a bad one by referring only to a mathematical model: organisational, pedagogical and cultural aspects of the whole decision process which leads to a given decision also contribute to its quality and success..." [Roy, 1990b, p. 27]. Thus, it becomes impossible to found the validity of a procedure either on a notion of *approximation* (i.e. discovering pre-existing truths) or on a mathematical property of *convergence* (i.e. does the decision automatically lead, in a finite number of steps, to the optimum a*?). The final solution is more like a "creation" than a discovery. In *multiple criteria decision aid* (MCDA) [Roy, 1985, 1990b, 1990c], the principal aim is not to discover a solution, but to construct or create something which is viewed as liable to help "an actor taking part in a decision process either to shape, and/or to argue, and/or to transform his preferences, or to make a decision in conformity with his goals" (*constructive or creative approach*) [Roy, 1990a, p.28].

Finally, we can conclude that the validity of a given procedure depends on two main factors:

- mathematical properties which make it conform to given requirements;
- the way it is used and integrated in a decision process.

4.3   Some Basic Definitions

Since in literature there is not a unanimous agreement on the definitions that should apply to some basic concepts of multicriteria decision aid, in this section, we will indicate clearly the definitions we are going to use in this book.

Criterion: a criterion is the basis for evaluation. It is a function that associates each action with a number ("number" in this contexts means any type of criterion score, quantitative, qualitative, stochastic or fuzzy) indicating its desirability according to consequences related to the same point of view. In formal terms, criterion g is a function defined on the set A of potential actions so that the comparison of the two numbers g(a) and g(b) allows us to describe and/or argue the result of the comparison of **a** and **b** relative to the point of view underlying the definition of g. Criteria may be further classified into goals (or targets) and objectives.

Goal: a goal (synonymous with target) is something that can be either achieved or not (e.g. increasing sales of a product by at least 10%). If a goal cannot be or is unlikely to be achieved, it may be converted to an objective.

Objective: an objective is something to be pursued to a maximum extent. (e.g. a company wants to maximise its level of profits). An objective generally indicates the direction of change desired.

Attribute: an attribute is a measure that indicates whether goals have been met or not, given a particular decision that provides a means of evaluating the levels of various objectives.

Constraint: a constraint is a limit on the values that attributes and decision variables may assume and can or cannot be stated mathematically.

Image of A in the criteria space: each action $a \in A$ may be represented in R^n by a point with co-ordinates $g_1(a), g_2(a), ..., g_n(a)$. The set of points thus obtained is the image of A in the criteria space.

Ideal point: the point which, in the criteria space, has co-ordinates

$$\max_{a \in A} g_1(a), \max_{a \in A} g_2(a), ..., \max_{a \in A} g_n(a) \qquad (4.2)$$

is called ideal (or utopia) point (assuming the higher the better principle). It is clear that if an element of A has this point as its image, this element is the best action, because it maximises simultaneously all the criteria. A concept similar to the ideal alternative, its mirror image, the anti-ideal can be defined as the action that minimises simultaneously all the criteria considered (the worst action).

Dominance: an action **a** dominates an action **b** if **a** is at least as good as **b** for all the criteria taken into consideration, and much better than **b** for at least one criterion.

Efficient solution: an action $a \in A$ is efficient if there is no action **b** in A which dominates **a**. The concept of efficiency can easily be illustrated graphically (see Figure 4.1). Alternative C performs better than B in all respects and hence C is preferred to B. The same can be said for B compared with A. Thus only C and D are efficient alternatives. It has to be noted that efficiency does not imply that every efficient solution is necessarily to be preferred above every non-efficient solution; e.g., the non-efficient alternatives A and B are preferable to the efficient alternative D if the second criterion would receive a high priority compared to the first criterion [Nijkamp et al., 1990].

The principle that inefficient solutions may be ignored could give rise to different problems:
• The assumption that all the relevant criteria have been identified needs to be accepted. If relevant criteria are omitted, there are potential opportunity costs associated with assuming that it is safe to ignore dominated alternatives.

• The assumption that only one alternative considered the best has to be identified needs to be accepted. Since the "second best" may have been eliminated during the technical screening, if more than one action has to be found, the elimination of the "inefficient" action may result in an opportunity loss (one has to note that if the best action is removed from the set of feasible alternatives, then the second best becomes a member of the non dominated set) [Bogetoft & Pruzan, 1991].

• A third problem is connected to the question: how relevant are "irrelevant" alternatives? [Zeleny, 1982] Arrow's axiom of "the independence of irrelevant alternatives" [Arrow, 1951] states that the choice made in a given set of alternatives A depends only on the ordering made with respect to the alternatives in that set. Alternatives outside A (irrelevant since the choice must be made within A) should not effect the choice inside A. Unfortunately, empirical experience does not generally support this axiom; thus to exclude some actions already inside A can have even less justification.

• Finally, a dominated action may be slightly worst than an efficient action, if indifference and/or preference thresholds are used (see section 4.7.2.1), then the two actions could present an indifference relation.

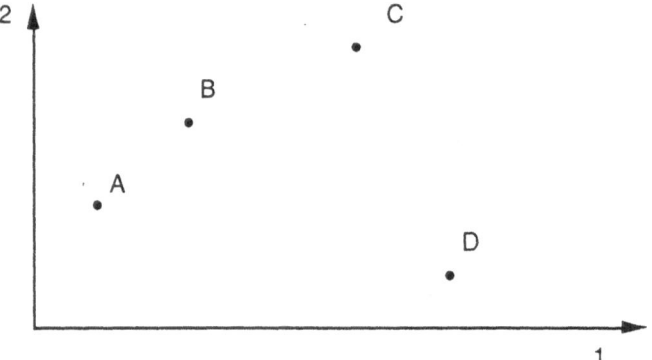

Figure 4.1   Efficiency in a Two-Dimensional Case

A multicriteria evaluation problem can be tackled by means of the following steps:

(1)   definition and structuring of the problem;
(2)   definition of a set of evaluation criteria;
(3)   choice between discrete and continuous methods;
(4)   identification of the preference system of the decision-maker(s) (interactive procedures or weighting systems);

(5)   choice of an aggregation procedure.

These steps will briefly be examined in the following sections.

4.4   Problem Definition

The results of any decision model depend on the available information; since this information may assume different forms, it is useful that decision models can take them into account. But, it has to be noted that this available information depends on the problem definition phase. According to systems methodology, the problem definition process may be synthesised in the following hierarchy of epistemological levels of systems [Cavallo, 1979; Klir, 1969]:
* *source systems* (all possible data that may be gathered),
* *data systems* (measurement of all variables),
* *generative systems* (relations among variables),
* *structure systems* (simplified representation of the whole system),
* *metasystems* (changements in time and space of the structure system).

The following considerations may be made:
(1)   the information used as input for decision models may be handled and structured in different ways, therefore a subjective component is always present;
(2)   it is generally accepted by the scientific community that the performance of each particular MCDA method is context dependent and that the model must fit real-world problems as close as possible, but it has to be noted that in decision problems, in a last analysis, "reality" is the result of the problem definition phase. Furthermore, it is not exact to assume that between the problem definition phase and the choice of the model there is a pure relationship of causality, in fact often the two phases are deeply interrelated (e.g. sometimes the same problem may be formulated both in continuous or discrete terms, and then the problem definition phase depends on this choice). Interesting suggestions for problem definition and structuring can be found in Bana e Costa [1993], Guimaraes Pereira et al. [1994], Norese [1991] and Ostanello [1990].

4.5   Definition of a Set of Evaluation Criteria

In formulating a set of evaluation criteria, two main tendencies can be distinguished. On one hand, one may wish to build a decision model as close as possible to the real world problem; this may increase the number of evaluation criteria to a level such that its applicability becomes almost impossible. On the other hand, one may wish to use a small number of criteria so that the model is simpler and faster to use; this may lead to an oversimplification of the model used.

According to Bouyssou [1990], a family of criteria should have two important qualities:

(1) the "legibility", i.e. the family should contain a sufficiently small number of criteria so as to be a discussion basis allowing the analyst to assess inter-criteria information necessary for the implementation of an aggregation procedure;

(2) the "operationality", i.e. the family should be considered by all actors as a sound basis for the continuation of the decision-aid study.

In addition, a family of criteria must also satisfy a number of technical properties, leading to the concept of a *consistent family of criteria* [Roy, 1985].

4.6   Choice Between Continuous and Discrete Methods

The number of alternatives may vary between 1, any discrete number and infinity. The problems with only one alternative is essentially a 0-1 choice system, in which a choice has to be made between the status quo and a new situation. Given the complexity of decision-making problems, it is not always possible to define the set A a priori. It may happen that the definition of A is progressively elaborated during the course of the decision aid procedure. It is also possible to distinguish between the cases where A is

globalised: each element of A excludes any other;

fragmented: the decision procedure's results involve combinations of several elements of A [Vincke, 1992].

4.6.1   Interactive Multiobjective Programming Methods

The main characteristic of multiobjective programming methods is that the feasible alternatives are only implicitly defined, so that in principle, their number is infinite. This problem has been analysed by various authors who have developed a large number of theorems and algorithms [Steuer, 1986].

A multiple objective program is a problem which aims at finding a vector $x \in R^n$, which satisfies a set of constraints (obeying eventual integrality conditions) and maximises a set of objective functions. A multiple objective program is said to be linear if the functions representing both the constraints and the objectives depend linearly on x.

Formally, let us consider a linear multiobjective problem (MOLP):

$$\max q \tag{4.3}$$

$$\text{subject to } q \in Q = \{f(x) : x \in X\} \tag{4.4}$$

$$\text{where } f(x) = \{f_1(x), f_2(x), \dots, f_k(x)\} = Cx \tag{4.5}$$

$$\text{and } X = \{x : Ax \leq b, x \geq 0\} \tag{4.6}$$

where Cx expresses linear relationships between m policy variables and k policy objectives, and $Ax \leq b$ expresses n linear constraints.

A multiobjective programming method can be divided into two phases:
- generation of the set of efficient solutions (also called Pareto set),
- exploration of this set in order to find a "compromise solution".

The set of efficient solution vectors is denoted by Q^E.

In MOLP problems the set of efficient solutions is called the efficient frontier (see Figure 4.2). Generally, in most continuous procedures, only extreme points solutions are used.

In a continuous framework, efficient alternatives can be generated in three different ways:

(a)   In theory it has been shown [Geoffrion, 1972] that all the efficient solutions can be generated by solving the following scalar maximum problem:

$$\max m^t q \tag{4.7}$$

$$\text{subject to } q \in Q \tag{4.8}$$

There are two problems in finding all efficient solutions:

first of all a scalar parametric problem cannot be solved in the open interval $\{m : m > 0\}$, but it can be solved in the closed interval $\{m : m \geq \varepsilon, \varepsilon > 0\}$. Thus the solutions are found if ε is "small enough", but "small enough" depends on the problem and thus cannot be specified a priori. This means that the choice of the ε can cause the exclusion of some efficient solutions. Moreover one may not find a vector m which gives a unique solution of the problem considered [Korhonen & Wallenius, 1988].

66

(b)   A second way of generating efficient alternatives consists in a systematic variation of side-conditions. Thus, by optimising one objective function under constraints on the other objective functions, which have to be varied in a systematic way, efficient alternatives are obtained.

(c) A third way of generating efficient alternatives consists in a systematic variation of weights in an objective function defining the "distance" between an appropriately chosen reference point and a feasible solution.

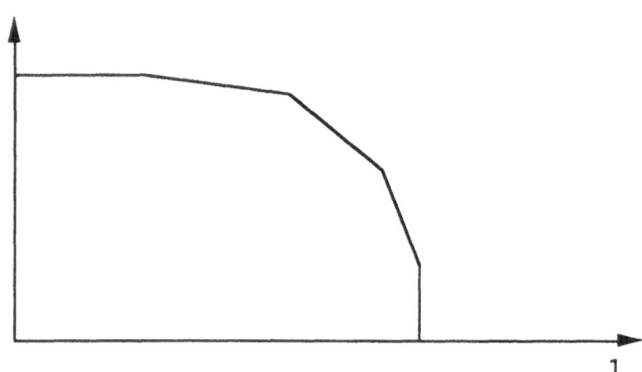

Figure 4.2   Efficiency Frontier in a Two-Dimension Case

In conclusion, we have to note that there are two problems in operating solely with efficient extreme point solutions:

•   in problems of a realistic size, the number of efficient extreme point solutions is very large;

•   the decision-maker is not necessarily satisfied with an extreme point solution as an approximation to the most preferred solution.

The above problems create a need to explore the efficient frontier carefully. This is done by means of interactive procedures.

Interactive procedures, since they are useful in order to explore the efficient frontier, are typical of those models which operate in a continuous context, although they exist also in some discrete models. Unlike the criterion weighting process, the closest collaboration between decision-maker and analyst occurs in interactive procedures when the model has already been put to use. The interaction process can rely on the following types of phases:

•   search of a candidate for a compromise solution,

•   communication to the decision-maker,

•   reaction of the decision-maker.

Vanderpooten and Vincke [1989] distinguish two main groups of interactive procedures:
* *search-oriented interactive procedures,*
* *learning oriented interactive procedures.*

In the first group, it is assumed that the decision-maker has a clear stable preference structure in his mind and that he acts in a consistent way with this structure. This approach may be criticised from different points of view:
* decision-makers do not have full knowledge about their preferences initially;
* preferences change as a function of time;
* the implicit assumption underlying this approach is that rationality coincides with consistency, but psychological researches have demonstrated that intransitivities and incomparabilities are a normal feature of human beings.

In the learning oriented interactive procedures, according to MCDA philosophy, no assumption is made on decision-maker's preference structure and consistency. But, also here some problems arise:
* since the preference structure is not assumed to remain stable and no consistency is required [Vanderpooten & Vincke, 1989], the information supplied by the decision-maker can not be considered as valid throughout the whole iterative process; therefore which information has to be taken into account?
* since all the hypotheses of the search procedures are relaxed, no mathematical convergence exists. This is not a problem if it is assumed that the decision-maker stops the procedure when a "satisfactory compromise solution" is found; but the reasons why the decision-maker decides to stop the procedure may be different (time, psychological climate, even colours of the screen!).

Nijkamp and Voogd [1985] have expressed the following opinion with regard to interactive procedures: "interactive procedures have several benefits. They provide information to the decision committee in a stepwise way, they can easily be included in a dynamic decision environment, they lead to an active role of all participants involved, and a priori specification of preferences or weights is not strictly necessary, although they can be inferred ex post. A limitation of this approach is that the final solution can depend on the procedure followed and especially on the starting solution. In addition for several continuous evaluation methods there is no guarantee that the compromise solution can be obtained within a finite number of interactive cycles, unless it is assumed that the decision committee is acting in a consistent way".

4.6.2 Discrete Methods

A discrete multicriteria problem may be described in the following way: A is a finite set of n feasible actions (or alternatives); m is the number of different points of view or evaluation criteria g_i i=1, 2, ... , m considered relevant in a decision problem, where $g_i : A \rightarrow R, \forall$ i=1,2, ... , m is a real valued function representing the i-th criterion according to a non decreasing preference, while the action **a** is evaluated to be better than action **b** (a, b∈ A) according to the i-th point of view iff $g_i(a) > g_i(b)$.

In this way a decision problem may be represented in a tabular or matrix form. Given the sets A (of alternatives) and G (of evaluation criteria) and assuming the existence of n alternatives and m criteria, it is possible to build an n x m matrix P called evaluation or impact matrix whose typical element p_{ij} (i=1, 2 , ... , m; j=1, 2 , ... , n) represents the evaluation of the j-th alternative by means of the i-th criterion. The impact matrix may include quantitative, qualitative or both types of information. In the terminology introduced by Vansnick [1990], this type of decision problem can be defined *"model A.A.E."* (Alternatives, Attributes, Evaluators).

This general description implies that evaluation problems may lead to different kinds of outcomes; for instance, some methods only aim at determining a set of acceptable alternative solutions, while other methods aim at the selection of one ultimate alternative. Thus there is a range of multicriteria problem formulations, which may take one of the following forms [Roy, 1985]:

(α) the aim is to identify one and only one final alternative;

(β) the aim is the assignment of each action to an appropriate predefined category according to what one wants it to become afterwards (for instance, acceptance, rejection or delay for additional information);

(γ) the aim is to rank all feasible actions according to a total or partial preorder;

(δ) the aim is to describe relevant alternatives and their consequences.

In discrete MCDA problems, there are several procedures aiming at obtaining decision-maker's priorities in the form of weights. In synthesis, it is possible to distinguish two main approaches [Nijkamp et al., 1990]:

- *direct estimation of weights* (trade-off method, rating method, ranking method, verbal statements, paired comparisons);
- *indirect estimation of weights* (weights based on previous choices, weights based on a ranking of alternatives, interactive estimation of weights).

All these methods present different features in terms of time needed, complexity, transparency, etc., and therefore their performance depends on the specific problem faced, but in general, the weighting of criteria is open to criticism for the following reasons:

1) in an abstract hypothesis it is accepted that the decision-maker has a clear idea in his mind of his own scale of preferences, and that he is capable of expressing these clearly without contradiction, while concretely, the logic of the choice assumes a reduction of the confusion inevitably present in the mind of the decision-maker when he is about to face a problem [Munda, 1993]. This initial state of confusion has been proved empirically; in fact, in experimental research it has been noted that, regarding the psychological climate surrounding the decision, there are substantially three phases to the decision process:

(1) initial disorientation,
(2) re-orientation process,
(3) solution [Riva, 1983].

As a consequence, it may occur that the decision-maker, even after weighting the different criteria with precision, is supplied by a method with results which do not satisfy him because the solution obtained is no longer in accordance with his scale of real preferences. Many scientists maintain that the decision-maker is not satisfied because he is inconsistent in his preferences, thereby forgetting that an inconsistent decision-maker represents the reality to which the mathematical model must be adapted; therefore, unless we admit explicitly that the model is a purely formal one, the irrationality is in the model, not in the decision-maker!

2) Another important point concerns the interpretation of the meaning of the weights supplied by the decision-maker. Weights are used to refer either to "scaling factors" or to "coefficients of importance". Such a meaning is strictly connected to an important feature of multicriteria methods, i.e. the concept of *compensation* [Vansnick, 1986; Vincke, 1989]. In the case of non-compensatory methods, the intercriteria information required is a relation of relative importance between coalitions of criteria. Such a concept of relative importance is often

translated into numbers called weights[2]. In the case of compensatory methods (e.g., weighted sum) the weights have to be considered as scaling factors and then their meaning is of a trade-off ratio.

The above considerations imply that given an aggregation procedure, there should be consistency between the aggregation procedure used and the questions asked to the decision-maker in order to elicitate a set of weights. Otherwise one runs the risk of combining weighting techniques with aggregation models with which they are not theoretically compatible. Even if the weights are elicitated in a well-defined and consistent procedure, it may happen that such weights are not precisely determined. In such cases two solutions are possible:

- sensitivity analysis aiming at verifying (by means of different vectors of weights) the robustness and stability of the results obtained with the initial vector of weights. But as it has been noted "one has to recognise that this procedure does not directly and specifically deal with imprecise weights, being only, a way of bypassing the problem [Bana e Costa, 1990b]";
- procedures aiming at directly facing situations of poor weighting information (e.g., Outweigh Analysis, Bana e Costa, 1990b; Regime Analysis, Nijkamp et al., 1990). A common problem of this kind of procedures is that a lot of assumptions need to be made for their correct axiomatization.

3) In cases of group decisions, it is often an impossible task to establish a weighting of the different criteria which satisfies all the decision-makers. For this reason, Roy after the experiment regarding the building of the Paris underground network, in ELECTRE IV, decided to eliminate the weighting of criteria.

4.7   Choice of an Aggregation Procedure

4.7.1   Utility Theory Approaches

4.7.1.1   Multiattribute Utility Theory (MAUT)

This theory is based on the following hypothesis: in any decision problem there exists a *real valued function U* defined on the set A of feasible actions, which the decision maker wishes, consciously or not, to examine. This function aggregates the different criteria taken into consideration, so that the problem can be formulated as

---

[2] In ELECTRE methods, the concept of importance of a criterion is taken into account by means of
- its weight (importance coefficient), and
- its veto threshold.

max U(g$_i$(a)) : a∈ A   (4.9)

where $U(g_i(a))$ is a utility function aggregating the m criteria (therefore a multicriteria problem is replaced by a monocriterion one). The role of the analyst is to determine this function. The most usual functions are the *linear* [3] or the *multiplicative* form [Bell et al., 1977; Keeney & Raiffa, 1976].

The main underlying assumption of this approach is the identification of human rationality with consistency. More analytically, it is assumed that the preference relations of the decision maker are a complete preorder, i.e.:
- all the states are comparable,
- transitivity of the preference relation,
- transitivity of the indifference relation.

Therefore, in the *"complete transitive comparability axiom"*, preferences can be modelled by means of two binary relations I and P having the following properties:
- I (indifference): reflexive, symmetric and transitive,
- P (strict preference): irreflexive, asymmetric and transitive.

Simon [1983] notes that humans have at their disposal neither the facts nor the consistent structure of values nor the reasoning power needed to apply the principles of utility theory. For instance, the assumption that human behaviour is transitive, can give rise to the famous Luce's paradox [Luce, 1956]: if T$_i$ represents a cup of coffee containing i milligrams of sugar, it is obvious that nobody comparing cups of coffee will sense a difference of one milligram, but he will generally express a preference between a cup of coffee with a lot of sugar and one with no sugar; this contradicts transitivity of the indifference relation (if Mark loves Judy and Judy loves John, according to transitivity Mark should love John! [Yu, 1990]).

From a mathematical point of view, a very important MAUT concept is that of trade-off or marginal rate of substitution, since it formalises the *notion of compensation*.

---

[3] The linear form can be used only if the condition of preference independence holds. A subset of attributes A is preferentially independent of A^C (the complement of A) only iff any conditional preference among elements of A, holding all elements of A^C fixed, remain the same, regardless of the levels at which A^C are held. The sets of objectives F$_1$, F$_2$,...,F$_n$ are mutually preferentially independent if every subset of A is preferentially independent of its complement A^C.

Let g_r be a given criterion, a first and approximate definition of a trade-off ratio is the necessary increase on criterion g_r to compensate the loss of one unit in criterion g_i

$$Sg_{ir} \text{ such that } (g_1, ..., g_i-1,..., g_r+Sg_{ir},...,g_n) \text{ I } (g_1, ..., g_i,...,g_r,...,g_n). \qquad (4.10)$$

A more precise definition can be given whenever $U(g)$ can be differentiated:

$$Sg_{ir} = \frac{\dfrac{\partial U(g)}{\partial g_i}}{\dfrac{\partial U(g)}{\partial g_r}} \qquad (4.11)$$

Since MAUT allows complete compensation among all the criteria, it is defined a complete compensatory model. One has to note that if environmental and economic criteria are taken into consideration, complete compensation implicitly means complete substitution between man-made capital and natural capital.

To summarise, this approach has the following consequences:

(1) a multicriteria problem is replaced by a monocriterion one;

(2) a problem which is not well stated is replaced by a well stated one;

(3) a problem including incomparable actions is replaced by a problem for which all actions are comparable;

(4) the evaluations of the different criteria are completely compensatory;

(5) indifference and/or preference thresholds cannot be used.

These consequences imply that sometimes the nature of the decision problem at hand may be completely changed.

4.7.1.2 The Analytic Hierarchy Process (AHP)

This method has been developed by Saaty [1980]. The AHP structures the decision problem in levels which correspond to one's understanding of the situation: goals, criteria, sub-criteria, and alternatives. By breaking the problem into levels, the decision-maker can focus on smaller sets of decisions. The AHP is based on 4 main axioms:

(1) Given any two alternatives (or sub-criteria), the decision-maker is able to provide a pairwise comparison of these alternatives under any criterion on a ratio scale which is reciprocal.

(2) When comparing any two alternatives, the decision-maker never judges one to be infinitely better than another under any criterion.

(3) One can formulate the decision problem as a hierarchy.

(4) All criteria and alternatives which impact a decision-problem are represented in the hierarchy.

The above axioms describe the two basic tasks in the AHP: formulating and solving problem as a hierarchy, and eliciting judgements in the form of pairwise comparisons [Harker, 1989]. The elicitation of priorities for a given set of alternatives under a given criterion involves the completion of a nxn matrix, where n is the number of alternatives under consideration. However, since the comparisons are assumed to be reciprocal, one needs to answer only n(n-1)/2 of the comparisons. Saaty proposed an enginvector approach for the estimation of the weights from a matrix of pairwise comparisons. The enginvector approach is a theoretically and practically proven method for estimating the weights. The enginvector also has an intuitive interpretation in that it is an averaging of all possible ways of thinking about a given set of alternatives.

After estimating the weights, the decision-maker is also provided with a measure of the inconsistency of the given pairwise comparisons. It is important to note that the AHP does not require decision-makers to be consistent but, rather, provides a measure of inconsistency as well as a method to reduce this measure if it is deemed to be too high.

After generating a set of weights for each alternative under any criterion, the overall priority of the alternatives is computed by means of a linear, additive function.

4.7.1.3   Utility Theory Approaches to Qualitative Multicriteria Evaluation

In multicriteria evaluation theory, a clear distinction is made between quantitative and qualitative methods. Essentially, there are two approaches for dealing with qualitative information: a direct and an indirect one [Nijkamp et al., 1990]. In the *direct approach*, qualitative information is used directly in a qualitative evaluation method; in the *indirect approach*, qualitative information is first transformed into cardinal, while next, one of the existing quantitative methods is used. Cardinalisation is especially attractive in the case of available information of a "mixed type" (both qualitative and quantitative data). In this case, the application of a direct method would usually imply that only the qualitative contents of all available (quantitative and qualitative) information is used, which would give rise to an inefficient use of this. In the indirect approach, this loss of information is avoided; the question is of course, whether there is a sufficient basis for the application of a certain cardinalisation scheme. Two examples of cardinalisation of a qualitative evaluation matrix are the expected value method [Rietveld, 1984, 1989] and multidimensional scaling techniques [Kruskal, 1964; Keller & Wansbeek, 1983; Nijkamp, 1979].

(1)   In the expected value method ordinal criterion scores are replaced by quantitative scores by using a transformation procedure aiming at deriving the centroid of a convex polyhedral set S consistent with the underlying ordinal information. In more formal terms:

let us assume that the ordinal information on the scores of n alternatives according to a given criterion is

$$p_1 \leq p_2 \leq \ldots \leq p_n \qquad (4.12)$$

then by standardising the p_i's according to the following formula

$$\sum_{i=1}^{n} p_i^{\beta} = 1 \quad (\beta > 0) \qquad (4.13)$$

and assuming $\beta \to \infty$, the constraint set S reads:

$$S = \left\{ 0 \leq p_1 \leq p_2 \leq \ldots \leq p_n = 1 \right\} \qquad (4.14)$$

If we assume that the p_i's are uniformly distributed on S, the probability density function shown in (4.15) results.

$$
\begin{cases}
(n-1)! & \text{if } 0 \leq p_1 \leq 1 \\
& \text{if } p_1 \leq p_2 \leq 1 \\
& \quad . \\
& \quad . \\
& \quad . \\
& p_{n-2} \leq p_{n-1} \leq 1 \\
0 & \text{elsewhere}
\end{cases} \qquad (4.15)
$$

On the basis of this function the expected values of criterion scores can be derived and it is not difficult to prove that the expected value is identical to the centroid of the polyhedron S. One has to note that this statistical distribution gives rise to a linear cardinalisation curve. If one wants to derive a concave structure, it can be proved that the probability density function in (4.15) has to be rewritten in a more complex way. However, by means of integration, Rietveld [1989] proves that a series of precise expected values can be obtained. These results can again be interpreted in terms of the centroid of a certain polyhedron. Weighted summation can now be used to rank the alternatives.

(2) A completely different approach to cardinalisation is the use of multidimensional scaling techniques. Such techniques aim at transforming qualitative data input into a cardinal output of lower dimensionality. In a sense, a scaling technique may be regarded as a kind of qualitative principal component analysis. It is clear that several concepts from multidimensional scaling analysis may also be applicable to ordinal multiple criteria problems. For instance, one may use a scaling technique in order to transform a qualitative evaluation matrix into a cardinal matrix with lower dimensionality. Then the cardinal configuration of the initial qualitative matrix provides a metric picture of the Euclidean distances both between the alternatives and between the effects. This is a normal standard operation. A limitation of this elegant but complex evaluation approach is that it requires a sufficient number of degrees of freedom to allow a multidimensional scaling. This implies that unless a sufficient number of evaluation criteria are used, no consistent scaling results can be obtained [Nijkamp et al., 1990].

An example of a multicriteria method that may use mixed information is the so-called REGIME method; this method is based on pairwise comparison operations; from this point of view it has something in common with outranking methods. However, it is based on a weighted linear additive model, thus it may be classified as a utility based method [Hinloopen & Nijkamp, 1990; Nijkamp et al., 1990]. Its point of departure is an ordinal evaluation matrix and an ordinal weight vector. Given the ordinal nature of the evaluation criteria, by means of pairwise comparison of alternatives, no attention is paid to the size of the difference between the impacts of alternatives; it is only the sign of the difference that is taken into account. Ordinal weights are interpreted as originating from unknown quantitative weights. A set S is defined containing the whole set of quantitative weights that conform to the qualitative priority information. In some cases the sign will be the same for the whole set S, and the alternatives can be ranked accordingly. In other cases the sign of the pairwise comparison cannot be determined unambiguously. This difficulty is circumvented by partitioning the set of feasible weights so that for each subset of weights a definite conclusion can be drawn about the sign of the pairwise comparison. The distribution of the weights within S is assumed to be uniform and therefore the relative sizes of the subsets of S can be interpreted as the probability that alternative **a** is preferred to alternative **b**. Probabilities are then aggregated to produce an overall rating of the alternatives, based on a success index or success score.

Let us present a simple illustrative example of a mixed-information based problem. Suppose that there are 3 possibilities for improving the transportation system in a region, viz. highway construction, a road/bus system and a new train (railroad) system (for more details on this problem see Janssen, 1992). Each of these 3 alternatives will be judged on the basis of 5 criteria, viz. costs, travel time, capacity, NO_x emissions and landscape impacts. Some of these impacts are quantitative, but others are qualitative in nature.

The mixed impact matrix related to the above problem is illustrated in Table 4.1. By applying the regime method to the problem described above the matrix of relative pairwise success indices presented in Table 4.2 is obtained. From this table it is clear that the train option is the most preferable alternative, followed by road/bus and highway. The value 1.00 in the comparison between train and highway alternatives indicates that for this comparison no added value is to be expected from a measurement of these criteria on a higher measurement scale. The probability that, given the ordinal information on travel time and landscape, the road/bus alternative ranks higher than the highway alternative equals 70%.

		Alternatives			
Criteria	Units	Highway	Road/bus	Train	Weights
Costs	mln gld	200	250	400	++
Travel Time	---/+++	+++	++	+	+
Capacity	mln km/year	20	30	40	+++
NO_x Emissions	ton/year	1000	750	100	+++
Landscape	---/+++	---	---	-	+

The ---/+++ scale is interpreted as an ordinal scale.

Table 4.1 Ordinal Evaluation Matrix of a Transportation Problem

	Highway	Road/bus	Train
Highway	-	0.30	0.00
Road/bus	0.70	-	0.01
Train	1.00	0.99	-

Table 4.2 Pairwise Success Indices Obtained by Means of REGIME

Another interesting method able to tackle mixed information is the EVAMIX method [Voogd, 1983]. The EVAMIX approach concerns the construction of two measures: one only dealing with the ordinal criteria and the other one dealing with the quantitative criteria. By making various assumptions about standardisation and aggregation, several methods can be defined by which an appraisal score for each alternative can be calculated. The most important assumptions behind the EVAMIX approach concern the definition of the various standardisation functions (at least three different techniques can be distinguished). Other assumptions concern the weights for the ordinal and cardinal criteria, and finally the additive relationship of the overall dominance measure. The global structure of the EVAMIX method is synthesised in Figure 4.3.

A problem, connected to all multicriteria methods that try to take mixed information into account, but that is particular evident in the EVAMIX approach is the problem of equivalence of the used procedures in order to standardise the various evaluations of the performance of alternatives according to different criteria. Mathematical techniques to deal with this problem will be shown in Chapters 6 and 7. However, it has to be noted that what we will take into consideration is qualitative information represented by means of fuzzy sets and not by means of ordinal information.

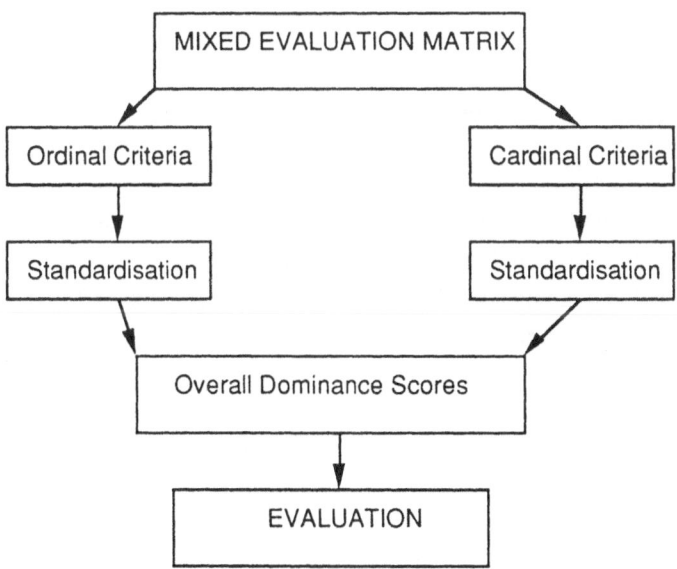

Figure 4.3   Scheme of EVAMIX

4.7.2.1   Outranking Methods

This approach is based on what Roy calls *"fundamental partial comparability axiom"* [Roy, 1985]; according to this axiom, preferences can be modelled by means of four binary relations **I** (indifference), **P** (strict preference), **Q** (large preference), and **R** (incomparability).

By means of the relation of large preference, all the other relations can be obtained:

- a**P**b ⇔ a**Q**b and not b**Q**a
- a**I**b ⇔ a**Q**b and b**Q**a
- a**R**b ⇔ not a**Q**b and not b**Q**a

In order to avoid giving a discriminating role to differences that are scarcely significant, indifference and preference threshold are introduced.

A criterion g is a *pseudo-criterion* if there exist two threshold functions q(g) (indifference threshold) and s(g) (threshold of presumed preference) such as g(a)≥g(b):

(1)   g(a)>g(b)+s(g(b)) ⇔ a**P**b

(2)   g(b)+q(g(b))<g(a)≤g(b)+s(g(b)) ⇔ a**Q**b

(3)   g(b)≤g(a)≤g(b)+q(g(b)) ⇒ a**I**b

In order to avoid some inconsistencies, the threshold functions must satisfy the following conditions:

$$g≥g' ⇒ g+q(g)≥g'+q(g') \text{ and } g+s(g)≥g'+s(g') \qquad (4.16)$$

$$s(g)≥q(g) \text{ for all g.} \qquad (4.17)$$

A graphical representation of a pseudo-criterion can be found in Figure 4.4.

This kind of modelling procedure may present a serious lack of stability. One may indeed imagine examples where small variations of scores lead to important variations of pseudo-orders, and thus to various conflicting prescriptions. Such undesirable discontinuities make a sensitivity analysis (or robustness analysis) necessary [Perny & Roy, 1992].

According to the presence of thresholds, the following classification holds:

a *semi-criterion* is a pseudo-criterion such as s(g)=q(g);

a *pre-criterion* is a pseudo-criterion such as q(g)=0 or is not defined;

a *true-criterion* is a pseudo-criterion such as s(g)=q(g)=0.

The true-criterion is the classical model used in decision theory, the first two using the notion of threshold enable modelling situations for which indifference is not transitive.

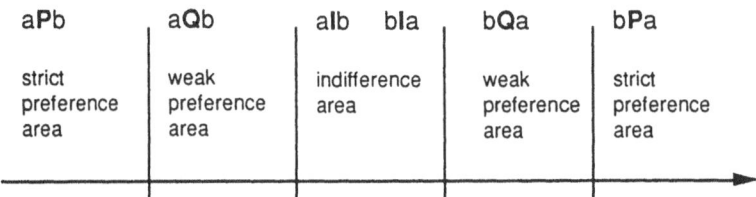

Figure 4.4   Structure of a Pseudo-Criterion

The concept of partial comparability is the base of the so-called *"outranking methods"*. These are based on the understanding that, in general, in multicriteria problems the dominance relation is poor because it is based on a consensus of points of view. Thus, an action **a** outranks an action **b** only if **a** is at least as "good" as **b** on all the criteria considered. The concept behind the outranking methods is that the enrichment of the dominance relation can be done only if realistic information is available; so there is a formal structure between the dominance relation which is too weak and the utility functions complete preorder. By using outranking methods some incomparable actions become comparable because realistic information exists, but other actions remain, nevertheless, incomparable.

Briefly, these models consist of aggregating the criteria into a partial binary relation aSb (outranking relation, see (4.19)) and then of the "exploitation" of this relation; each of these two steps may be treated in a number of ways according to the problem formulation and the particular case considered. A common property of these methods is that they are completely or partially non compensatory. For an extensive discussion of the concept of compensation in MCDA see [Bouyssou, 1986; Vansnick, 1986]. One should note that to limit the possibility of compensation between environmental and economic criteria allows to operationalize a strong sustainability philosophy.

The first outranking methods were the ELECTRE methods, developed by Roy and his collaborators [Roy, 1968, 1977, 1990b; Roy & Bertier, 1973]. Nowadays different MCDA methods belonging to this family are present in the literature; for example, PROMETHEE [Brans et al., 1986], a sub-family belonging to the so-called PCCA (Pairwise Criterion Comparison Approach) [Matarazzo, 1991]: MAPPAC [Matarazzo, 1986], PRAGMA [Matarazzo, 1988], CARTESIA

[Giarlotta, 1990], IDRA [Greco, 1992]. Outranking approaches to qualitative multicriteria evaluation are ORESTE [Pastijn & Leysen, 1989] and MELCHIOR [Leclercq, 1984].

As an example, we shall briefly present ELECTRE 1 when it is applied to a family of criteria without discrimination thresholds, in ELECTRE 1, the proposition **aSb** is accepted if the concordant coalition $C(aSb) = \{g_i \in G : g_i(a) \geq g_i(b)\}$ is sufficiently important (condition of concordance) and if on the other criteria the difference $g_i(b) - g_i(a)$ are not too large (condition of non-discordance). The importance of a coalition is represented by the sum of the weights (w_i) of the criteria belonging to that coalition. Thus the index $c(a, b)$ defined by:

$$c(a, b) = \frac{\sum\limits_{i \in C(aSb)} w_i}{\sum\limits_{i=1}^{m} w_i} \qquad (4.18)$$

represents the relative importance of $C(aSb)$ among the set of all criteria. Whether or not $C(aSb)$ is sufficiently important is then judged comparing $c(a, b)$ to a threshold $s \geq 1/2$ called concordance threshold.

In order to determine which differences on the discordant criteria are judged too large, a veto threshold v_i (that may vary with g_i) is defined on each criterion in such a way that the existence of a discordant criterion such as $g_i(b) - g_i(a) \geq v_i$ prohibits acceptance of aSb whatever the value $c(a, b)$. Thus in ELECTRE 1:

$$aSb \Leftrightarrow \left[c(a, b) \geq s \text{ and } g_i(b) - g_i(a) < v_i \ \forall \ i \notin C(aSb) \right] \qquad (4.19)$$

Examples of other possible aggregation procedures are the lexicographic model, the ideal point approaches and the aspiration levels models.

4.7.2.2 Outranking Approaches to Qualitative Multicriteria Evaluation

MELCHIOR [Leclercq, 1984] has been developed for the case in which an importance relation on the criteria is at hand, but there is no desire to quantify it by weights, thus only ordinal importance relations are used. In this method a family of m pseudo-criteria is taken into consideration. The basic idea is to say that **a** outranks **b** if the criteria which are unfavourable to the latter assertion are "hidden" by those which are in its favour and if no criterion g_j exists such as $g_j(b) > g_j(a) + v_j$, where v_j is a veto threshold.

In order to define "favourable" and "unfavourable" criteria to a given relation of outranking, and what is meant by criteria "hidden" by other criteria, several definitions are proposed by the author. By choosing two combinations of definitions, one stricter than the other, one obtains a strong and a weak outranking relations which are in turn exploited as in the ELECTRE IV method. Vincke [1992] notes that the choice of combinations of definitions is not arbitrary, but a set of coherent combinations can be obtained.

The ORESTE method [Pastijn & Leysen, 1989] has been developed for the case in which both criterion weights and scores are ordinal in nature. This method can be divided into two main phases:

phase 1: construction of a global (complete) weak order on the set A of feasible actions;

phase 2: construction of an incomplete preference structure on A, after an indifference and conflict analysis.

The actions are ranked by means of a methodology similar to the famous additive rank-ordinal method of Borda [Moulin, 1988]. ORESTE is very discriminatory about conflictual actions and it allows incomparability relations. The global preference relation obtained is transitive in nature. This method is not a purely ordinal one, since some quantitative aspects in the treatment of the data are introduced.

D'Avignon and Vincke [1988] have developed a method for the treatment of situations in which the criterion scores are under the form of probability distributions. The main characteristics of this method are :

- it is based on pairwise comparison of actions;
- it preserves, for as long as possible, the distributional feature of the criterion scores and it does not replace them from the start by a unique number (e.g. mean or median);
- the notion of relative importance of criteria is incorporated in the method.

The various steps of the method have points in common with the PROMETHEE and ELECTRE methods.

As one can see, different interesting outranking approaches to different aspects of the presence of qualitative information in MCDA have been developed, however none of them tackles the issue of "mixed information".

4.7.3 The Lexicographic Model

This is the model used to put in order the words in a dictionary, the first letter playing the role of the first criterion, the second letter, the second criterion, and so on. To use the model, the decision maker must give a total strict order on the criteria:

$$1>2>... >i > ... > m \qquad\qquad (4.20)$$

where g_1 would be the most important criterion and g_m the least important. In the lexicographic model, all actions are first ranked by means of the first criterion, then if some indifferent actions exist, these are further explored by means of the second criterion, and so on. It has to be noted that this procedure is completely non compensatory. Lexicographic orders usually lead to a straightforward selection of the most preferred alternative, however, most of the information collected on alternatives will not play a role in the choice process. The lexicographic model can be combined with other different approaches e. g. multiobjective programming methods [Isermann, 1982].

4.7.4 Ideal Point Approaches

Ackoff [1978] writes:"An ultimately desired outcome is called an "ideal". If one formulates a problem in terms of approaching an ideal solution, one minimises the changes of overlooking relevant consequences in decision making. Seeking the ideal is the best way to open and stimulate the mind to creative activity". Briefly, the philosophy underlying the multicriteria methods based on ideal point concepts can be synthesised as follows [Yu, 1973, 1985; Zeleny, 1974, 1982]. Multicriteria problems are characterised by conflicts because of the perceived absence of a prominent alternative; therefore, the only way to dissolve conflicts is to find or invent the ideal point. The only way to decrease the intensity of conflict is to find or generate alternatives which are as close as possible to the ideal point. Coombs [1958] assumes that there is an ideal level of attributes for objects of choice and that the decision-maker's utilities decrease monotonically on both sides of this ideal point. He shows that probabilities of choice depend on whether compared alternatives lie on the same side of the ideal or on the other. The ideal point procedures are characterised by the following axiom of choice: *alternatives that are closer to the ideal are preferred to those that are farther away. To be as close as possible to the perceived ideal is the rationale of human choice.*

A concept similar to the ideal alternative, its mirror image, the anti-ideal can be defined on any properly bounded set of feasible alternatives. The question is, do humans strive to be as close as possible to this ideal or as far away as possible from the anti-ideal? Zeleny [1982] writes: "our answer-both. As a matter of fact we propose that humans are capable of switching between the two regimes according to the given circumstances of the decision process...Naturally, the compromise set based on the ideal is not identical with the compromise set based on the anti-ideal. This fact can be used in further reducing the set of available solutions by considering the intersection of the two compromises".

One of the traditional ideal point approaches is to compute the "distance" of each action from the ideal point and then rank them in terms of their closeness to the ideal. One problem related to this approach is that each action is considered completely independent from the set of all the other actions; the assumption is that humans compare each action with the ideal rather than among themselves. On the contrary, in MCDA literature it is assumed that a desirable property is that the ranking of actions must also be a function of the relationships among the different actions taken into consideration. One possible way to do so is to recompute the distances of all the remaining alternatives each time one is removed from the set of feasible actions. "As all alternatives are compared with the ideal, those farthest away are removed from further consideration. There are many important consequences of such partial decisions. First, whenever an alternative is removed from consideration there could be a shift in a maximum attainable score to the next lower feasible level. Thus the ideal alternative can be displaced closer to the feasible set. Consequently, the removal of any alternative affects the ranking of the remaining alternatives in terms of their closeness to the ideal. Similarly, addition of a new alternative could displace the ideal farther away by raising the attainable levels of attributes [Zeleny, 1982]". One weak element in this procedure is that the ranking of alternatives is affected by changing the ideal point, which is an artificial action. On the contrary, it should be desirable that the ranking is a function of real actions.

4.7.5 Aspiration Levels Models

Aspiration levels (or goals) express the decision-maker's ideas about the desired outcomes of the decision in terms of a certain level to be aimed at for each criterion or objective. There is a close link between the concept of aspiration level and the theory of satisfying behaviour [Simon, 1983].

The usual way in which aspiration levels are treated is by means of goal programming [Spronk, 1981]. An advantage of goal programming is that it always provides a solution, even if none of the goals are realisable, provided that the feasible region is non empty. This is possible by using deviational variables, which show whether the goals are attained or not. In the latter case, they measure the distance between the realised and aspired levels.

An approach that can be regarded as a generalisation of goal programming and ideal point techniques is the "achievement scalarizing functions" method [Wierzbicki, 1982]. The main idea is constructing a mathematical basis for satisfying decision making by introducing the wishes of the decision-maker as a basic a priori information in the form of aspiration levels (reference points). Achievement scalarizing functions can be considered as a modification of traditional utility functions.

4.8   Cost-Benefit Analysis and Multicriteria Evaluation

In this Chapter, we have shown that multicriteria methods provide a flexible way of dealing with qualitative environmental effects of decisions. However, this does not mean that multicriteria evaluation is a panacea which can be used in all circumstances without difficulties. It has its own problems, and some of these problems have also been illustrated. In this section, a comparison of the key characteristics of cost-benefit analysis and multicriteria evaluation will be carried out on the base of 13 comparison criteria, that is: rationality assumed, mathematical axiomatization, economic axiomatization, problem structuring, analytical cost, alternatives taken into account, evaluation criteria, preference system, democratic basis, aggregation procedures, comprehensiveness, empirical applicability, transparency and sustainability.

1) Rationality Assumed

Cost-benefit analysis is based on the Neo-Classical maximisation premise on behaviour stating that rational decisions coincide with utility maximisation. Consistency is also considered an important characteristic of rationality; as a consequence, the preference structure is assumed to hold only the preference and the indifference relations, both relations are considered of a complete transitive type.

Multicriteria decision theory is based on different types of rationality according to the models used. However, the only one similar, to some extent, to the concept of rationality assumed in Neo-Classical theory is MAUT. In general, models based on satisfying behaviour, bounded and procedural rationality principles are considered. In outranking methods even a constructive decision support is assumed; no transitivity of indifference relations is implied, and the relation of incomparability is also used.

2) Mathematical Axiomatization

The mathematical axiomatization of cost-benefit analysis is complete, since in a monocriterion analysis it is possible to discover a precise optimal solution, if it exists. Furthermore, CBA is based on standard investment criteria such as NPV or IRR. However, it is not always clear which investment criterion should be used. If the NPV is used, since it can be considered an additive utility function, the condition of preference independence should always hold.

Since a multicriteria problem is by definition mathematically ill-structured, i.e. it has no objective solution, a complete mathematical axiomatization of multicriteria evaluation is very difficult. This is also the most important cause of the flourishing of a lot of different theories and models. This is a weak point of the multicriteria approach. However, a way of bypassing this problem is to indicate clearly the axiomatic system underlying any method and to list the set of properties considered desirable.

3) Economic Axiomatization

CBA is completely consistent with the maximisation and the weighting premises of Neo-Classical economics. According to the scarcity principle, the most efficient allocation of resources is considered the most important objective of economic analysis. On the contrary, in multicriteria evaluation no one-dimensional monetary performance indicator is considered. Efficiency is not considered the only aim of the analysis but many different conflictual heterogeneous points of view are considered. If Neo-Classical economics is considered the best (or the only) economic paradigm, then CBA is the most consistent appraisal tool with this economic theory. If, above all in the light of environmental problems, one considers it desirable to revise the premises of Neo-Classical economics, then multicriteria evaluation may be an interesting decision tool. Since multicriteria methods allow one to take into account conflictual, multidimensional, incommensurable and uncertain effects of decisions, they can be considered perfectly consistent with the methodological foundations of ecological economics.

4) Problem Structuring

In CBA the main effort consists in trying to apply the right valuation techniques to transform everything into money terms. As shown in Chapter 3 the monetisation of environmental costs and benefits is very difficult. Different valuation techniques exist, and it is not always clear which is the best technique to apply in a certain real-world problem; thus the problem of choosing the right method for the right problem is not only connected to multicriteria evaluation, but it is also present in CBA. In a multicriteria application, no standard problem structuring exists; even the same problem can be structured in completely different ways. This increases the subjective component of the analysis, but on the other hand, it makes the multicriteria methodology much more flexible than a CBA one.

5) Analytical Cost

The application of CBA may be quite time consuming; this is mainly because the application of valuation techniques may be very complex. The application of multicriteria evaluation is normally less time consuming than CBA; very poor information (ordinal or even linguistic) can be used.

6) Alternatives Taken Into Account

CBA is limited to discrete problems. Even only one alternative can be evaluated versus a base alternative. Arrow's axiom of independence of irrelevant alternatives is always respected since the degree of attractiveness of each single alternative is independent from all other actions.

In multicriteria decision theory, any finite or even infinite number of alternatives can be taken into account. With respect to the axiom of independence of irrelevant alternatives, some methods (e.g. MAUT) respect it; while the results provided by other methods (e.g. outranking methods) are also a function of the set of alternatives considered.

7) Evaluation Criteria

In CBA it is necessary to identify all costs and benefits and then their correct transformation into monetary values. In a multicriteria problem, a reasonably large number of criteria, reflecting completely different points of view, can be taken into account. The criterion scores can be quantitative or qualitative. The possibility of taken into account qualitative consequences of decisions is very important for environmental problems where, normally, intangibles are often present.

8) Preference System

In CBA economic votes expressed on the market are in principle taken into account. However, one has to note that it is not possible to escape from value judgements since, if no weighting system is used, the assumption that current distribution of income is optimal needs to be accepted. Multicriteria evaluation explicitly recognises that the identification of the preference system of the decision-maker(s) is a very important step of the overall analysis. We think that in an evaluation exercise, the presence of a subjective component has to be accepted. The advantage of multicriteria methods is that the subjectivity is made explicit.

9) Democratic Basis

Some equity problems arise since normally willingness to pay measures consider preferences of the higher income groups more important than the lower ones and future generations are not considered. In multicriteria evaluation, generally the decision-makers to whom one can ask the weights or with whom the interaction can be carried out are political authorities in charge of a given decision. From this point of view multicriteria decision theory can be considered more elitaristic than CBA. However, equity problems can explicitly be considered in three different ways:

- as specific evaluation criteria,
- by means of different sets of weights,
- by integrating multicriteria methods with conflict resolution techniques.

10) Aggregation Procedures

In CBA only utility based models of a complete type are used, as a consequence, complete compensation is assumed. In multicriteria decision theory, various aggregation procedures exist; models with completely different properties can be used. This makes multicriteria evaluation more flexible but also more confusing since a method has to be chosen and the final results are very sensitive to this step. However, we have already noted that this problem of "method uncertainty" is also present in CBA, since different valuation techniques exist.

11) Comprehensiveness

Equity and environmental aspects can hardly be incorporated in a cost-benefit analysis. Multicriteria evaluation is by definition multidimensional in its nature. If it is accepted that efficiency, equity and sustainability are different objectives in economic theory, these can more easily be tackled in a multicriteria analysis than in a CBA one.

12) Transparency

Since in CBA everything is translated into money terms and then aggregated in a compensatory fashion, the possibility of a complete understanding of all the profiles taken into consideration is quite low. However, the possibility of presenting the background data can increase the transparency of a given application.

In multicriteria evaluation, all the multidimensional profiles of the problem are clearly shown in the original scales of measurement. From an operational point of view, the great transparency allowed in multicriteria models is one of its most important factors of success. One should remember that the communication with the public is a fundamental aspect of environmental management problems.

13)   Sustainability

Is CBA consistent with a goal of sustainable development? Since cost-benefit analysis is based on the compensation model, the only definition of sustainable development that can be operationalized is the weak sustainability concept.

If the Pearce and Turner definition of sustainable development is considered, this could be operationalized by means of cost-benefit analysis only if the government receives sufficient shadow projects to offset environmental damages, so that across a portfolio of public investments, net environmental damage is zero. However, besides the aggregation problems inherent in this definition of sustainability already discussed in Chapter 2, there are problems here, both in measuring environmental impacts and in designing shadow projects which fully compensate (see Chapter 3).

Is multicriteria evaluation consistent with sustainable development? Since multicriteria evaluation is multidimensional in nature, it allows to take into account economy-environment interactions. According to the aggregation procedure chosen, weak or strong sustainability concepts can be operationalized. Also uncertain environmental impacts can be taken into account. Issues, particularly important in a regional context, such as multiple use and inter-regional spatial links and trade-offs can be tackled by means of multicriteria evaluation (given the assumption of a second best world).

To choose between CBA and MCDA is not a trivial point, we could say that it is a meta-multicriteria problem. Van Pelt [1993] maintains that CBA is more attractive than MCDA from a methodological point of view, while MCDA is preferable from an empirical point of view. However, we have shown that from a pure methodological point of view, CBA has the advantage of being consistent with Neo-Classical economics; if other economic paradigms are considered this is not an advantage anymore. Furthermore, some assumptions underlying the rationality of CBA (e.g., utility maximisation, transitivity of indifference and preference relations) can also be disputable. Differently from van Pelt, we also think that equity issues can more easily be incorporated in a MCDA exercise than in a CBA one.

One should note that regarding environmental problems, CBA and MCDA can be considered as competitive methods only if all environmental consequences of decisions can be correctly transformed in monetary values; but we have seen in Chapter 3 that this is very difficult. Thus we can say that, given the presence of unpriced environmental impacts, often multicriteria evaluation is the only possible approach (if fuzzy uncertainty is considered too, the use of MCDA is even more advisable). However when monetary values are present, CBA can be used as a criterion of MCDA in a successful way.

Following the interesting methodology of decision trees to choose between CBA and MCDA proposed by van Pelt [1993], our view can be schematically illustrated in Figure 4.5. Like van Pelt, the *nature of criteria* (we use efficiency, equity and sustainability) and the *type of information available on attributes* are used as critical variables for the selection of a method. Also cost-effectiveness analysis (CEA) is considered.

As one can see, if environmental attributes cannot be translated into money terms the use of MCDA is desirable. However, if efficiency attributes are completely or partially in money terms, CBA or CEA can be integrated with MCDA.

Finally, one should note that the results of a cost-benefit analysis depend on:
- identification of costs and benefits and their monetisation
- choice of a social rate of discount
- choice of a time horizon
- construction of a one-dimensional indicator bringing together all the benefits and costs.

On the other hand, the results of a multicriteria analysis depend on:
- available data
- structured information
- chosen aggregation method
- decision-maker's preferences.

This means that when an attempt is made to model a real world situation, the presence of a certain subjective component appears to be an inevitable phenomenon. In general, this is a desirable feature, in fact when a model without any creative, personal or subjective influence of a model designer is used, this is inevitably characterised by a certain rigidity which prevents it adhering completely to the situation modelled. This could make it necessary to "force reality" because in the end the tendency will be to make reality fit the model. The use of models with characteristics of subjectivity or of subjectivism, depends in the latter analysis *on the ability and ethical behaviour of the researcher constructing the model*. It is important to remember this above all, when MCDA or CBA methods are used to "justify" or "defend" political decisions.

90

Because of these deep uncertainties present in evaluation methods, it is a case of "post-normal science". In such cases, the traditional subject-specialty expertise is inadequate for peer-review of quality. Quality assurance therefore requires "extended peer communities", which include all those with a stake in the issue who are prepared to dialogue. One should note the consistency between this notion of partecipation of post-normal science and the concept of quality of the decision process in a constructive framework theorised by multicriteria decision aid. Moreover, multicriteria decision aid and post-normal science share the principles of incommensurability and value conflicts [see O'Connor et al., 1995].

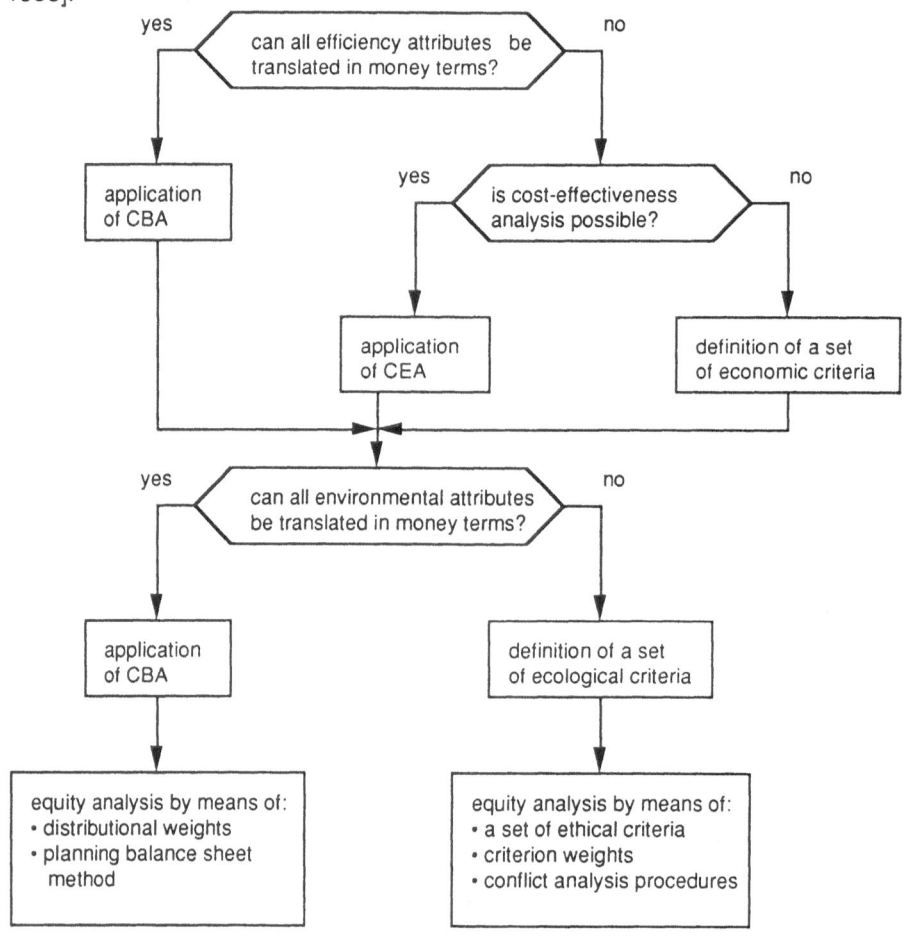

Figure 4.5   A Possible Way to Choose Between CBA and MCDA

PART B

MULTICRITERIA EVALUATION IN A FUZZY ENVIRONMENT

CHAPTER 5

FUZZY UNCERTAINTY IN DECISION MODELS

5.1   Introduction

In this Chapter, we will illustrate some basic concepts of fuzzy set theory. Given the enormous amount of literature on this field, only the concepts necessary to the understanding of the rest of the book will be presented.

Ideally the information available for a decision model should be precise, certain, exhaustive and unequivocal. But in reality, it is often necessary to use information which does not have these characteristics and to face the uncertainty of a stochastic and/or fuzzy nature present in the data. Therefore, the combinations of the different levels of measurement with the different types of uncertainty have to be taken into consideration (see Figures 5.1 and 5.2).

	Quantitative Information	Qualitative Information
Stochastic		
Non-Stochastic		

Figure 5.1   Possible Combinations of Information Measurement Levels and Stochastic Uncertainty

These types of information have been deeply studied in measurement theory and in probability theory and in statistics; but some open problems still remain, above all regarding the representation of ordinal information in evaluation and decision problems.

	Quantitative Information	Qualitative Information
Fuzzy		
Non-Fuzzy		

Figure 5.2   Possible Combinations of Information Measurement Levels and Fuzzy Uncertainty

Fuzzy sets as formulated by Zadeh [1965] are based on the simple idea of introducing a degree of membership of an element with respect to some sets. *Fuzzy nominal information* is the base of the whole fuzzy set theory, since it considers all cases between 0 (non membership) and 1 (complete membership), and it is represented by means of the membership functions.

Let us assume that the symbol U means the entire set (Universe of discourse). In classical set theory, given a subset A of U, each element $x \in U$ satisfies the condition: either x belongs to A, or x does not belong to A. The subset A is represented by a function $f_A : U \rightarrow \{0, 1\}$:

$$f_A(x) = \begin{cases} 1 & \text{if } x \in A \\ 0 & \text{if } x \notin A \end{cases} \qquad (5.1)$$

The function f_A is called a characteristic function of the set A. Fuzzy sets are then introduced by generalising the characteristic function f_A. Let U again be a universe of discourse. Let $x \in U$. Then a fuzzy set A in U is a set of ordered pairs

$$\{[x, \mu_A(x)]\}, \quad \forall x \in U \qquad (5.2)$$

where $\mu_A : U \rightarrow M$ is a membership function which maps $x \in U$ into $\mu_A(x)$ in a totally ordered set M (called the membership set) and $\mu_A(x)$ indicates the grade of membership of x in A. Generally, the membership set is restricted to the closed interval [0, 1]. A fuzzy set is completely determined by its membership function. For $0 < \mu_A(x) < 1$, x belongs to A only to a certain degree; thus there is ambiguity in determining whether or not x belongs to A. The physical meaning is that a gradual instead of an abrupt transition from membership to non-membership is taken into account. A classical example is that of age. Let U be the set of all non-negative integers. Let us take into consideration the primary terms young and old. These terms can be considered the label of two fuzzy sets A and B. No doubt the ages 6 or 10 are young, whereas the ages 30 or 40 are less young. Thus it is possible to define a membership function μ_{Ayoung} showing the degree of compatibility of the age x to the concept of young.

It is indispensable however to clarify here a point of fundamental importance: the use of membership functions [Munda et al., 1993a]. *Membership functions* constitute the essential basis on which the whole fuzzy set theory is built; they represent no doubt a brilliant idea which revolutionised traditional set theory, giving birth to a new mathematical field. But paradoxically, the membership functions constitute at the same time the strongest and the weakest point of the theory. Various scientists are sometimes sceptical with regard to fuzzy sets for the main reason that they consider these membership functions too subjective. Therefore, it is necessary to address the question, on which factors does such a subjectivity depend? Two essential factors may be distinguished here [Zimmermann, 1987]:

(1)   the context in which they are to be applied;

(2)   the method adopted in the building phase.

(1)   The membership functions depend on the semantic contents of the subjective category they represent and therefore they vary according to the context in which they are to be applied. Then the question is whether this feature is really a negative one. To answer this question, it is necessary to distinguish between subjectivity and subjectivism:

(a)   by *subjectivism* we mean the exaggeration of the subjective component, so that in this case the membership functions turn out to be simply the totally arbitrary fruit of whoever is modelling a specific situation; of course such an approach is methodologically wrong and leads to completely unreliable and untestable results, as it cannot be inter-subjectively validated on scientific grounds.

(b)   In general, when an attempt is made to model a real world situation, the presence of a certain subjective component appears to be an inevitable phenomenon. This can be illustrated as follows. When a classical model of a deterministic type could be used, without any creative, personal or subjective influence of a model designer, this is inevitably characterised by a certain rigidity which prevents it from adhering completely to the situation modelled. This could make it necessary to "force reality" because in the end the tendency will be to make reality fit the model; then we could fall into a clear case of irrationalism (an illustrative example of this is that of Hegel in the "Philosophy of nature" where in order to realise his dialectic project (thesis, antithesis, synthesis) he obliged nature to fit his model by eliminating two of the five continents!).

Given the previous observations, we claim that it may be desirable to have at one's disposal a model with good characteristics of flexibility of a fuzzy type in order to make it adhere more strictly to reality. Of course the use of membership functions with characteristics of subjectivity or of subjectivism, depends in the latter analysis on the ability of the researcher constructing the model. Therefore, in general, *the model itself constitutes a necessary element, but it is not sufficient; it must allow a degree of creativity that is necessary and sufficient for the direct action of the decision-maker who must be able, even at the last minute, to recover its direction and modify its orientation.*

(2)   Membership functions may be built in two different ways:
(a) *deductively*, with the use of formal models constructed according to specific hypotheses.
(b) *empirically*, with the use of two different methods:
(I) interpolating a finite number of degrees of membership,
(II) constructing a real model of a membership function and seeking to verify its empirical validity.

The empirical approach is more suitable for evaluation and decision models, mainly because it greatly reduces subjectivism.

Qualitative information can be represented by means of fuzzy sets in two different ways:
*   using linguistic variables,
*   using graphical procedures.

Both approaches will concisely be discussed in the next sections.

5.2   Linguistic Variables

In traditional mathematics, variables are assumed to be precise, but when we are dealing with our daily language, imprecision usually prevails. Intrinsically, daily languages cannot be precisely characterised on either the syntactic or semantic level. Therefore, a word in our daily languages can technically be regarded as a fuzzy set. Formally, a linguistic variable is represented by a quintuple $(X, T(x), U, G, M)$ [Leung, 1988; Zimmermann, 1986] where:
X is the name of the variable, e.g. age;
T(x) is the term set of X, finite or infinite, such as young, very young and so on, in an universe of discourse **U**. A primary term in $T(x)$ is a term whose meaning must be defined a priori, and which serves as a basis for the computation of the meaning of the non primary terms in $T(x)$;
G is a syntactic rule by which the non primary terms in the term set are generated. It is possible to use a context free grammar or a regular grammar; in G it is possible to find *primary terms, hedges* (not, very, more or less, etc.), *relations* (younger than, older than, etc.), *conjunctions* (e.g. and), and *disjunctions* (e.g. or).Thus computer implementation of such an approach

presents a high degree of complexity and generally requires artificial intelligence oriented languages;

M is a semantic rule which associates each term with its meaning (a fuzzy subset in U). Through M, a compatibility (membership) function $\mu : U \rightarrow [0, 1]$ is constructed (e.g. μ_{young} shows the degree to which a numerical age is compatible with the concept of young and equivalently μ_{young} may be viewed as the membership function of the fuzzy set young).

Therefore, a linguistic variable is a fuzzy variable whose values are fuzzy subsets in a universe of discourse. The base variable of the linguistic variable is a precise variable which takes an individual value in its domain, i.e. the universe of discourse U. The domain of the linguistic variable is the collection of all possible linguistic values, fuzzy sets defined in the same universe of discourse through the base variable. However, in some cases, the fuzzy set which is assigned to the fuzzy restriction may not have a numerically-valued base variable. In order to allow a formal analysis, a mathematical translation of such linguistic propositions is needed. This can be done by means of possibility theory[1] [Dubois & Prade, 1980].

If R(x) is a fuzzy restriction (a fuzzy restriction is a fuzzy relation which acts as a flexible constraint on the values that may be assigned to a variable), then the effect of F (a linguistic value) on X (base variable) can be expressed as

$$R(x) = F, \tag{5.3}$$

where X is a variable in U (Universe of discourse), F is a fuzzy set in U and R(x) is a fuzzy restriction imposed by F. Therefore, this fuzzy restriction may be expressed in a linguistic proposition:

$$P : X \text{ is } F \tag{5.4}$$

which generates a possibility distribution

$$\pi_X = F \tag{5.5}$$

Associated with the possibility distribution there is a possibility distribution function such that

---

[1] For an extensive discussion on the differences between probability and possibility see Dubois & Prade [1986, 1989].

Poss $(X=a) = \pi_X (a) = \mu_F (a)$ (5.6)

Therefore, we can conclude that fuzzy restrictions, possibility distributions and fuzzy sets are closely related.

In the qualitative information available for an evaluation or decision model, two different types of linguistic variables may be present:
(1) the meaning can be translated in a measure on an interval or ratio scale (quantitative base variable), e.g. age, distance, etc.;
(2) there is no meaning on an interval or ratio scale, and therefore the base variable is also qualitative in nature, e.g. appearance, comfort, beauty, etc.

Type 1. If linguistic variables whose meaning can be translated in a measure on an interval or ratio scale are present in a decision model, generally it is because of a lack of information or of the right instrument of measurement. Therefore, we have a qualitative evaluation of a variable that in theory could be measured on an interval or ratio scale. So it is reasonable to suppose that it is possible to transform the qualitative information into a quantitative one with a certain degree of precision. The parameters, necessary in this case, may be easily established, because this is a case of the so-called "*informational fuzziness*" depending mostly on the subjective culture of the person in charge of the evaluation. For example, the proposition "that man is tall" may have different meanings for different people, but everybody can easily indicate the "tolerance interval" of his own evaluation. Such a representation takes into account some "labels" of the term set, e.g. young, young and/or very young, etc., and the problem is to find which values of the base variable are compatible with these terms. Formally, in general we know that

Poss $(X=a) = \pi_X (a) = \mu_F (a)$ (5.7)

Thus, if for example, the linguistic proposition is "the distance is long", then

Poss $(distance=10 \text{ km}) = \pi_{distance} (10) = \mu_{long} (10) = 0.6$ (5.8)

indicates the possibility of 10 km being considered as a long distance. That is, the possibility distribution indicates whether or not the proposition "the distance is long" is possible and equivalently; it denotes the degree of compatibility of 10 km to the fuzzy set long. Therefore, the problem may be easily solved by establishing possibility distributions, e.g. $\pi_{distance}$ = long [Leung, 1988],

$$\pi_{distance}(x)=\mu_{long}(x)= \begin{cases} 0 & \text{if } x \leq \beta \\ 1 - e^{-k\left(\frac{x-\beta}{\beta}\right)^2} & \text{if } x > \beta \quad (k>0) \end{cases} \tag{5.9}$$

Type 2. In the case of linguistic variables with no meaning on an interval or ratio scale, the qualitative information does not depend on lack of information, but on the nature of the information that is essentially fuzzy (*intrinsic fuzziness*). Therefore, whereas in the other cases the stochastic representation and the fuzzy one may be competitive, in this case the fuzzy representation is the only one possible. For example, if linguistic propositions (like "pretty girl", "beautiful flower" or "quality of life") clearly have no quantitative base variable, how can we represent them? It seems that there is a set of hidden and fuzzy standards in one's mind in a justification for this type of concepts, but they are more than a human being can rationally handle simultaneously [Zimmermann & Zysno, 1983].

A first approach to this problem may be to try *to decompose the concept* that one wants to represent into a series of quantitative measurable variables. This approach presents two main problems, viz. the explication of the quantitative variables, and the aggregation procedure to be used.

A second approach is to define an *artificial quantitative base variable*, assuming that the real space is one-dimensional. The interval of the real space is chosen from [-1, 1], [0, 1], or [0, 10], etc., as desired and can be subdivided into a series of fuzzy sets representing linguistic values (e.g. very negative, moderately positive, very positive, etc.). Then, a link or mapping between quantitative (numerical) and qualitative (linguistic) values is established. This "*direct estimation*" approach has been criticised because of its lack of theoretical foundation. Recently, some psychologists [Norwich & Turksen, 1982, 1984; Wallsten et al., 1986] have developed a graded pair comparison procedure, which allows simultaneous testing of the necessary axioms, scaling of the responses in order to obtain memberships and tests of goodness of fit. Subsequently, empirical experiments have demonstrated a high level of similarity between membership values determined through graded pair comparison and direct magnitude estimation! Thus it seems that a theoretical justification for the quantification of the vague meanings of inexact linguistic terms by means of direct estimation can be established.

A third interesting approach can be the notion of *type 2 fuzzy set.* A type 2 fuzzy set is a fuzzy set whose membership values are fuzzy sets on [0, 1]. This corresponds to the case that the decision-maker is not able (or not willing) to characterise the grade of membership by an exact number, but gives an evaluation such as "the grade of membership is high, medium", etc. It is always possible to define a fuzzy set of type n=2,3,.. , if its membership function is a mapping from U to a set of fuzzy subsets of type n-1; therefore, it is possible that in order to reduce the fuzziness, many transformations are requested, thus diminishing drastically the computational efficiency of the algorithm.

Another possible approach has been developed above all in the field of psychological research [Hersh et al., 1979], called the "*yes-no paradigm*". In this approach, an element x of the universe of discourse U, is presented to the subject and he has to decide whether the element is a member of A, A being a fuzzy subset of U. The fraction of positive responses across replications (within or across subjects) is considered a measure of $\mu_A(x)$. It has been noted that the main problem of this approach is that it confounds fuzziness with response variability and that it can be interpreted as an indication that words have various, but nevertheless precise meanings to different people and/or different times.

5.3   Interactive Graphical Procedures Approach

Hesketh and others [1988] have noted that "by adapting traditional psychological measurement to deal with fuzzy concepts, new possibilities are open to both fields of enquiry". These authors have proposed a computerised graphic rating scale using linear membership functions and combining some rules of fuzzy set theory with others that are typical of Bayesian statistical theory in order to obtain as a compound index of the membership function the expected value of the re-scaled distributions (see Chapter 6). A traditional psychological procedure initially developed for the measurement of attitudes is the so-called "*semantic differential*". This approach requires an individual to describe a concept in terms of where it falls between bi-polar objective descriptions. Thurstone [1969], facing the problem of the meaning of attitudes, proposed a graphical representation of individual differences. Furthermore, he suggested that the range of opinions which a particular person is willing to endorse could also be represented graphically. "Using graphic representation, Thurstone demonstrated that an individual's opinion could be characterised in terms of three different measures, the range of opinions the individual is willing to endorse, their mean position on the scale, and the one opinion selected which best represented an attitude [Hesketh et al., 1988 p.429]". Hesketh and others in the fuzzy graphic rating scale, consider as a representation of a rater's perception, a point indicated by means of a mouse in a bi-polar scale and then the extension of the rating to the left or right (these extensions represent uncertainty inherent in this estimate). The main assumptions of this procedure are:

(1) the membership function takes the value 1 at the first point indicated by means of the mouse;

(2) the membership functions are linear;

(3) the union of two or more fuzzy ratings has a convex membership function;

(4) each fuzzy variable can be represented by means of its expected value, to be computed by rescaling its membership function.

In the light of these observations, different heuristic graphical procedures can be proposed in order to represent qualitative information as fuzzy sets. For example, if ordinal outcomes are available, the following procedure of cardinalization can be used. If according to an ordinal evaluation criterion, we know that $a_1 > a_2 > \ldots > a_n$, then on the basis of the principle of insufficient reasoning, it is possible to represent such alternatives by means of membership functions with the same shape, equally spaced and with all the intersections empty. This representation implies the assumption that the intensity of preference of each alternative to the one just before it in the ranking is constant, which again implies that the best alternative is much better than the worst. Furthermore, in case of mixed-information, very seldom it is possible to find quantitative evaluation criteria whose outcomes fall into the whole possible range of values (e.g. a car whose price is 0!); this implies that if such an assumption is accepted, the distance between actions on the ordinal criteria will be much greater than the one computed on the cardinal ones. For all these reasons, this representation appears not very plausible; therefore, it can be mainly used as a starting point to interact with a decision-maker. Then he can modify the *position*, the *shape* and the range of the membership functions according to his own evaluations. For this procedure, computer graphic assistance can be a very useful tool.

It has to be noted that the range defined by the decision-maker can be viewed as a measure of his uncertainty in giving the evaluation (the larger the range, the more uncertain the evaluation); this situation is represented by means of the intersections among the evaluations given (of course the larger the range, the more probable the intersections).

5.4 Fuzzy Numbers

Fuzzy quantitative information has been deeply studied in the fuzzy literature by means of the notion of fuzzy numbers [Dubois & Prade, 1980].

A **fuzzy number** is simply a fuzzy set in the real line and is completely defined by its membership function such as

$$\mu(x) : R \rightarrow [0, 1] \tag{5.10}$$

For computational purposes, in general this definition is restricted to those fuzzy numbers which are both normal and convex [Bonissone, 1982];

$$\text{normality: } \sup\{\mu(x)\} = 1 \quad \text{with } x \in R \tag{5.11}$$

$$\text{convexity: } \mu\{\lambda x_1 + (1 - \lambda)x_2\} \geq \min\{\mu(x_1), \mu(x_2)\}$$
$$\text{with } x \in R \text{ and } \lambda \in [0, 1] \tag{5.12}$$

The requirement of convexity implies that the points of the real line with the highest membership values are clustered around a given interval (or point). This fact allows one to easily understand the semantics of a fuzzy number by looking at its distribution and to associate it with a properly descriptive syntactic label (e.g. "approximately 10").

The requirement of normality implies that, among the points of the real line with the highest membership value, there exists at least one which is completely compatible with the predicate associated with the fuzzy number.

A standard normal convex *trapezoidal fuzzy number* (see Figure 5.3) can be characterised by a 4-tuple (a, b, α, δ) where [a, b] is the closed interval on which the membership function is equal to 1, α is the left-hand variation and δ is the right-hand variation. If only one point in the real line with $\mu(x) = 1$ exists, the fuzzy number is called a *triangular fuzzy number*.

A more general type of fuzzy number is the *L-R fuzzy number*; it is defined as follows:

$$\mu_A(x) = \begin{cases} F_L(x - m) / \alpha & \text{if } -\infty < x < m, \ \alpha > 0 \\ 1 & \text{if } x = m \\ F_R(x - m) / \delta & \text{if } m < x < +\infty, \ \delta > 0 \end{cases} \tag{5.13}$$

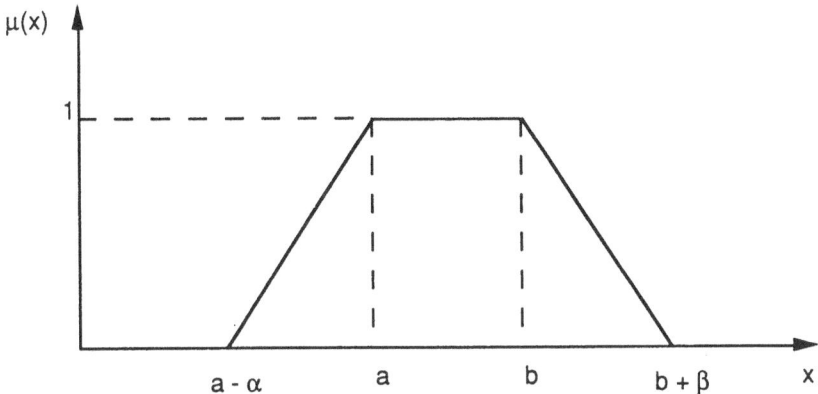

Figure 5.3   A Trapezoidal Fuzzy Number

where m, α, δ, are the "middle" value, the left-hand and the right-hand variation, respectively. $F_L(x)$ is a monotonically increasing membership function and $F_R(x)$, not necessarily symmetric to $F_L(x)$, is a monotonically decreasing function (see Figure 5.4). If a closed interval on which the membership function is equal to 1 exists, it is called a *flat fuzzy number*.

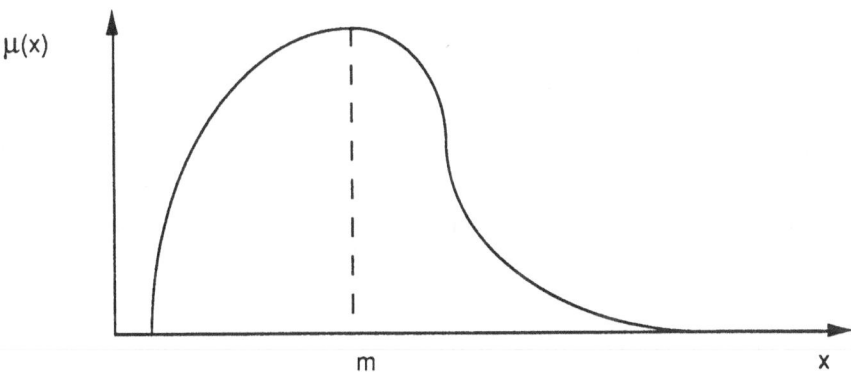

Figure 5.4   A L-R Fuzzy Number

5.5   Linguistic Hedges

The hedges which enter in the representation of linguistic values can be viewed as signs of various operations defined on the fuzzy sets of U. A linguistic hedge (or modifier) is an operation which modifies the meaning of a term or more general of a fuzzy set [Zadeh, 1972, 1973]. If A is a fuzzy set, then the modifier m generates the composite term B=m(A). Thus, if one knows what is meant by beautiful, then he also knows what is meant by "very beautiful", "not beautiful" and so on. Examples of mathematical models frequently used for modifiers are concentration (as an approximation of "very"), dilation (as an approximation of "more or less") and contrast intensification. "Very" acts as an intensifier, generating a subset of the set on which it operates. A simple operation which has this property is that of concentration. It is:

$$CON\ (A) = A^2 \tag{5.14}$$

and as an approximation it is possible to assume

$$\underline{very\ A} = CON\ (A) \tag{5.15}$$

Applying this operation to A results in a fuzzy subset of A such that the reduction in the magnitude of the grade of membership of x in A is relatively small for those x which have a high grade of membership in A and relatively large for the x with low membership.

The operation of dilation has an opposite effect to the one of concentration. By means of this operation the fuzziness of a fuzzy set is increased. It is:

$$DIL\ (A) = A^{1/2} \tag{5.16}$$

$$\underline{more\ or\ less\ A} = DIL\ (A) \tag{5.17}$$

The contrast intensification of a fuzzy set A is written INT (A) and is the set defined as

$$INT\ (A) = \begin{cases} CON\ (A) & \forall x : \mu_A(x) < 0.5 \\ DIL\ (A) & \forall x : \mu_A(x) \geq 0.5 \end{cases} \tag{5.18}$$

This has the property of increasing membership if greater than 0.5 and decreasing it if less than 0.5. It has to be noted that the level 0.5 is chosen arbitrarily.

5.6 Linguistic Operators

The basic operations with fuzzy sets are union, intersection and complement as in classical set theory. At the beginning of the development of fuzzy set theory, the min-max operations and the corresponding assumptions were the only ones used as intersection and union operators. Bellman and Giertz [1973] showed that they are the only operators which satisfy a series of axiomatic conditions, interpreting the intersection as "*logical and*" and the union as "*logical or*". In formal terms:

$$A \cup B: \mu_{A \cup B}(x) = max \{\mu_A(x), \mu_B(x)\} \tag{5.19}$$

or in logical symbols:

$$A \cup B: \mu_{A \cup B}(x) = \{x : x \in A \lor x \in B\} \tag{5.20}$$

$$A \cap B: \mu_{A \cap B}(x) = min \{\mu_A(x), \mu_B(x)\} \tag{5.21}$$

or in logical symbols

$$A \cap B: \mu_{A \cap B}(x) = \{x : x \in A \land x \in B\} \tag{5.22}$$

It has to be noted that these operators are non interactive, in the sense that there is no trade-off between their operands; thus a high value of one cannot compensate for a low value of another. A modification of A or B does not necessarily lead to a change in $A \cup B$ or $A \cap B$.

The complement of a fuzzy set A which is denoted by A^c is defined as follows:

$$A^c: \mu_{A^c}(x) = 1 - \mu_A(x) \tag{5.23}$$

Nowadays a lot of different operators have been defined in literature, imposing sets of new axiomatic conditions see Zimmermann [1986]. It has to be noted that many operators assume exactly the same value if the grades of membership are restricted to the values 0 or 1, while they lead to different results if that is no longer true. The variety of operators for the aggregation of fuzzy sets might make it difficult to decide which one to use in a specific model or situation; in fact, such a variety of operators makes the theory more flexible but at the same time more confusing in model formulations. Zimmermann [1986] has proposed some criteria on which the selection of appropriate operators may be based. But in a real-world problem, it is impossible to find an ideal-operator fitting all the requested criteria, and thus a certain degree of compromising needs to be made. In our view, it is not only important that the operators satisfy certain axioms or have certain formal qualities from a mathematical point of view, but the operators must also be appropriate models of real system behaviour, which can normally only be proven by empirical testing.

The problem of linguistic operators has been deeply studied also in psychology. But, to summarise the evidence regarding the and/or operations is inconclusive. Oden [1977] advocates the sum product rule, Thole and others [1979] favours the min rule for "and", and Hersh and Caramazza [1976] support the max rule for "or". Furthermore, given the various methods used to quantify the degrees of membership and the variety of experimental procedures used to elicit the judgements, it is impossible to determine whether the different results are really contradictory or just reflect the use of different methodologies and different situations. Finally Zimmermann and Zysno [1983] showed that subject's judgements are best represented as a combination of the two membership functions that do not fit any of the rules proposed for conjunction and disjunction. Their work clearly illustrates that in real life situations, information is sometimes combined according to an averaging rule that allows for compensation between the extreme alternatives. This result is also consistent with a large scale of empirical studies conducted by Anderson [1981, 1982] and his collaborators who demonstrates that the averaging combination rule is the preferred mode of information integration. Such an operator is:

$$\left(\prod_{m=1}^{M}\mu_m\right)^{(1-\gamma)}\left(1-\prod_{m=1}^{M}(1-\mu_m)\right)^{\gamma} \tag{5.24}$$

As shown by the authors, this is a convex combination of the product operator and the sum operator, which are respectively known as the algebraic representation of the intersection and the union. In this operator, μ_m is the normalised degree of membership and γ is a parameter indicating the degree of compensation[2].

---

[2] By compensation in the context of aggregation operators for fuzzy sets is meant the following: "Given that the degree of membership to the aggregated fuzzy set is $\mu_{Agg}(x_k) = f(\mu_A(x_k), \mu_B(x_k)) = k$, f is compensatory if $\mu_{Agg}(x_k) = k$ is obtainable for different $\mu_A(x_k)$ by a change in $\mu_B(x_k)$ [Zimmermann, 1986, p. 36]".

5.7 Level Sets and Decomposition

The concept of α-level sets is an important "means of communication" between ordinary sets and fuzzy sets. Since a fuzzy set lacks a precise boundary, then the identification of its element can be problematic. There is a need to impose a rule for determining unambiguous membership; one way is to require the elements of the corresponding ordinary set to possess a certain grade of membership in the fuzzy set. The concept of α-level sets is therefore formulated [Leung, 1988].

The α-level set of a fuzzy set A is the ordinary set A_α such as

$$A_\alpha = \{x : \mu_A(x) \geq \alpha\}, \quad \alpha \in [0, 1] \tag{5.25}$$

and it is defined by the characteristic function

$$f_{A\alpha}(x) = \begin{cases} 1 & \text{if } \mu_A(x) \geq \alpha \\ 0 & \text{if } \mu_A(x) < \alpha \end{cases} \tag{5.26}$$

By using the concept of α-level sets, the connection between ordinary sets and fuzzy sets can be established by the following relation (sometimes called "decomposition theorem"):

$$A = \bigcup_{\alpha \in [0,1]} \alpha \cdot A_\alpha \tag{5.27}$$

Thus any fuzzy set is a family of ordinary sets.

5.8 Measures of Fuzziness

Measures of fuzziness indicate the *degree of fuzziness* of a fuzzy set. Three main approaches can be distinguished [Zimmermann, 1986]:
- entropy measures [De Luca & Termini, 1972; Capocelli & De Luca, 1973],
- normalised distances [Kaufmann, 1975],
- degrees of distinction between the fuzzy set and its complement [Yager, 1979; Higashi & Klir, 1982].

Here, we only give a brief illustration of the entropy approach. The entropy of a fuzzy set A on a Universe X, with membership function $\mu_A(x)$, is given by

$$H(A) = \frac{1}{N} \sum_{x \in X} \ln(x) \tag{5.28}$$

where N is the number of elements in X (X being finite) and $\ln(x)$ is the *incertitude* of the evaluation along scale x given by

$$\ln(x) = -[\mu_A(x)\log_2\mu_A(x) + (1-\mu_A(x))\log_2(1-\mu_A(x))] \tag{5.29}$$

The entropy of a set A is 1 if for every x, $\mu_A(x)=0.5$ and is 0 if for every x, $\mu_A(x)=1$ or 0.

An interesting property is that

$$H(A) \geq H[INT(A)] \tag{5.30}$$

The philosophy underlying this concept is simple; if fuzzy sets are interpreted as a profile determined on a set of polar scales, it would be a very unclear evaluation if each scale response is in the middle. On the other hand, if each evaluation is extreme-oriented, then the evaluation is very clear.

5.9 Fuzzy Relations

Let X and Y be two universes of discourse, a fuzzy binary relation R in the Cartesian product XxY is a fuzzy set in XxY defined by the membership function

$$\mu_R : XxY \to [0, 1] \tag{5.31}$$

$$(x, y) \to \mu_R(x, y), \quad x \in X \text{ and } y \in Y \tag{5.32}$$

where the grade of membership $\mu_R(x, y)$ indicates the degree of relationship between x and y. If X=Y, the relation is called a fuzzy relation on X.

A fuzzy relation on X is said to be:

reflexive if	$\mu_R(x, x)=1$ and $\mu_R(x, y)<1$, $y \neq x$, $\forall x \in X$	(5.33)
symmetric if	$\mu_R(x, y)=\mu_R(y, x)$, $\forall(x, y) \in X$	(5.34)
asymmetric if	$\mu_R(x, y) \neq \mu_R(y, x)$, $\forall(x, y) \in X$	(5.35)
antireflexive if	$\mu_R(x, x)=0$, $\forall x \in X$	(5.36)
max-min transitive if	$\mu_R(x, z) \geq \max \min[\mu_R(x, y), \mu_R(y, z)]$, $\forall x,y,z \in X$	(5.37)
min-max transitive if	$\mu_R(x, z) \geq \min \max[\mu_R(x, y), \mu_R(y, z)]$, $\forall x,y,z \in X$	(5.38)

It is interesting to note that the entropy measure of fuzziness can be easily extended to the notion of fuzzy relations [Gupta et al., 1977]. Let R be a fuzzy relation on X, then its incertitude is given by

$$H(R) = -\frac{1}{n^2} \sum_x \sum_y \ln(x, y) \tag{5.39}$$

$$\ln(x, y) = -[\mu \log \mu + (1 - \mu)\log(1 - \mu)] \qquad \text{where} \tag{5.40}$$

$$\mu = \mu_R(x, y) \tag{5.41}$$

By using fuzzy relations, a formal analysis of imprecise linguistic relations like "distance x is much longer than distance y", "x is similar to y", etc, is allowed. For example, a fuzzy relation Γ representing the concept "much greater than" can be defined as follows:

$$\Gamma(x, y) = \begin{cases} f(x - y) & (x > y) \\ 0 & (x \leq y) \end{cases} \tag{5.42}$$

where f(x - y) is a monotone nondecreasing function defined on non-negative real numbers such as

$$x - y \leq 0 \Rightarrow f(x - y) = 0, \qquad \text{and} \tag{5.43}$$

$$x - y = \infty \Rightarrow f(x - y) = 1 \tag{5.44}$$

5.10 Multicriteria Evaluation in a Fuzzy Environment

A general multicriteria decision model (Δ) characterised by fuzzy information can be synthesised as follows:

$$\Delta = \{G, A, C, P, W, M\} \qquad \text{where}$$

G is the set of objectives, criteria or goals,
A is the set of feasible alternatives,
C is the set of constraints,
P is the set of relevant parameters,
W is the set of the subjective preferences of the decision-maker, and

M is the set of relevant membership functions.

Thus a fuzzy decision model is essentially characterised by a set of membership functions. These membership functions can be defined on one or more of the other components of the model, and therefore the degree of fuzziness of the model may vary accordingly.

In a decision problem it is possible to distinguish two main elements, available information and manipulation rules of this information. Accordingly, in multicriteria evaluation in a fuzzy environment two main classes can be distinguished [Munda, 1988]:

(1)   fuzzy manipulation rules of crisp information,
(2)   fuzzy manipulation rules of fuzzy information.

Since this second class is the main topic of the present book, we will survey more in deep some relevant approaches of this class of problems. As an example of fuzzy manipulation rules applied to crisp information, the fuzzy outranking concept will be briefly illustrated.

In Chapter 4 the notion of an outranking relation has been illustrated. A fuzzy outranking relation [Roy, 1977, 1978] is an outranking relation where with each ordered pair (a, b), a real number $\sigma(a, b)$, with $0 \leq \sigma(a, b) \leq 1$, is associated. $\sigma(a, b)$ is called the *credibility index* of the outranking aSb. Such an index characterises the degree of strength of the arguments bringing to the assertion aSb. The credibility index has been introduced into the method ELECTRE III, where it is opportunely combined with the concordance and discordance indices. An axiomatic characterisation of fuzzy outranking relations can be found in Perny & Roy [1992]; such authors also try to extend the notions of a fuzzy outranking relation to the case in which fuzzy criterion scores are present. However, as shown in Figure 5.5, the way by which fuzzy sets are compared is quite simple.

The starting point of decision models in a fuzzy environment was the Bellman-Zadeh [1970] model. Such authors assume membership functions defined on both goals and constraints, and define a decision in a fuzzy environment as "the confluence of goals and constraints", i.e. as the appropriate aggregation (intersection) of all the fuzzy sets.

In formal terms, let us assume a fuzzy goal represented by the membership function $\mu_G(a)$ and a fuzzy constraint represented by the membership function $\mu_C(a)$.

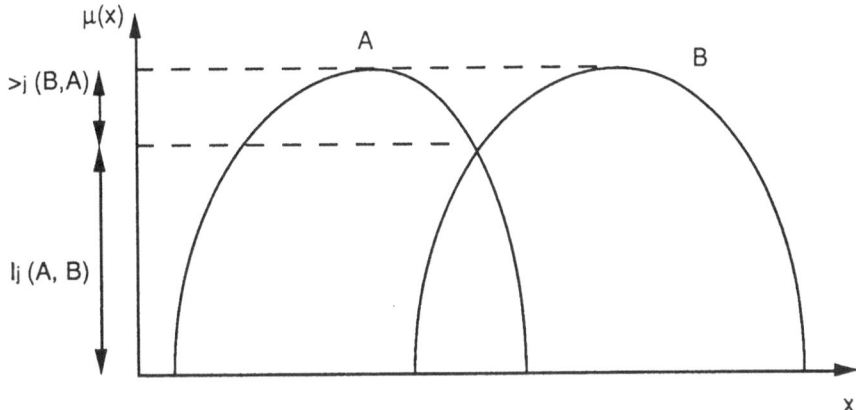

Figure 5.5   Fuzzy Binary Relation Derived from Fuzzy Scores

A decision is then defined by the following membership function

$$\mu_D(a)= \min \{\mu_G(a), \mu_C(a)\} \tag{5.45}$$

Then it is possible to find a fuzzy optimal decision, defined as

$$\max_{a} \min \{\mu_G(a), \mu_C(a)\} \tag{5.46}$$

Of course this definition implies the acceptance of the assumption that the intersection of fuzzy sets is defined by the min operator, however Bellman and Zadeh recognised that the min interpretation of the intersection might have to be modified depending on the context.

The Bellman-Zadeh model can easily be extended to the case where n goals and m constraints exist; in this case it is:

$$\mu_D(a)= \left(\bigwedge_{i=1}^{n} \mu_{G_i}(a) \right) \bigwedge \left(\bigwedge_{j=1}^{m} \mu_{C_j}(a) \right) \tag{5.47}$$

If a unique maximising decision belonging to this set exists, then it is a uniquely defined crisp decision which can be interpreted as the action which belongs to all fuzzy sets representing either constraints or goals with the highest possible degree of membership. However, it should be noted that in general a unique solution is not found but a set of "optimal" solutions exists. How to isolate a single action from this set is not a trivial problem. Furthermore, it has to be noted that sometimes between goals and constraints in the critical region Δ, some cases of fuzzy independence may occur. In this case the critical region cannot be found by means of formula (5.47).

The Bellman-Zadeh maxmin solution has been extensively used in fuzzy linear programming methods. Given the topic of the present work, in the following we will focus on discrete fuzzy multicriteria models. For an extensive overview of these methods see [Slowinski & Teghem, 1990; Zimmermann, 1987].

As shown in Chapter 4, a discrete multicriteria decision problem may be represented in a tabular or matrix form whose typical element p_{ij} represents the evaluation of the j-th alternative by means of the i-th criterion. In a fuzzy environment, either the evaluations associated with each alternative and the weights can be membership functions representing fuzzy numbers or linguistic variables. As a consequence of this, the manipulation rules normally used for crisp multicriteria methods have to be changed.

Yager [1978] assumes a simple model with fuzzy criterion scores and crisp weights. According to the Bellman-Zadeh model, a fuzzy set decision D is obtained by means of the intersection of all the membership functions representing the fuzzy criterion scores and then the maxmin rule is applied. The weights obtained by means of the well-known Analytic Hierarchy Process (AHP) by Saaty [1980], are expressed by exponentially weighting of the membership functions. This is justified in the light of the definition of the modifier "very"; this is normally defined as the squaring operation.

Laarhoven and Pedrycz [1983] extend Yager's model allowing for weights represented by triangular fuzzy numbers. Such weights are also derived by means of the AHP method, but since the matrices contain fuzzy numbers, the computations are far from simple.

Baas and Kwakernaak [1977] also propose a model able to deal with fuzzy criterion scores and fuzzy weights. In their model, first an index expressing the fuzzy rating of each alternative $\mu_R(r)$ is obtained as the fuzzy set that synthesises both the fuzzy criterion scores and the fuzzy weights. Then a ranking (in a complete preorder) of the alternatives is obtained by means of other two indices:

- $\mu_I(a_i)$ indicates the degree to which a_i is the best alternative and
- $\mu_{P_i}(p)$ indicates the degree of preferability of a_i over all other alternatives.

It has to be noted that such indices are obtained by means of standard fuzzy set rules using above all maxmin operations of intersections of the membership functions. Baas and Kwakernaak assume that all the membership functions are normal, piecewise continuous, differentiable, bounded, non-negative and that there exists a finite support. However, one should note that to determine explicitly the membership functions defined by the various indices is a problem which might not be solvable at all in general. Furthermore, sometimes the Baas and Kwakernaak method leads to results that are against our intuition. For example, Baldwin and Guild [1979] present the example illustrated in Figure 5.6 in which the Baas and Kwakernaak model would indicate a_1 as the best alternative, whereas intuition might regard alternative a_2 as preferable to a_1.

Approaches aimed at overcoming this problem inherent in the Baas and Kwakernaak model, have been proposed by Jain [1977] and Baldwin and Guild [1979]. However, it has to be noted that in Jain's model only the right side (decreasing parts) of the membership functions are taken into consideration. In the Baldwin and Guild model, a series of fuzzy relations that are supposed to represent differences of cardinal utilities are employed. Then standard max-min composition operations are used. In order to apply this model a number of assumptions have to be satisfied. In particular, for the fuzzy relations, the membership functions should be one sided and either strictly increasing or decreasing in each argument; for fuzzy ratings, the membership functions should be unimodal and with sides strictly increasing or decreasing, all the membership functions range over the whole interval [0, 1]. In any case, one should note that the computations involved are easy only when triangular fuzzy numbers are used.

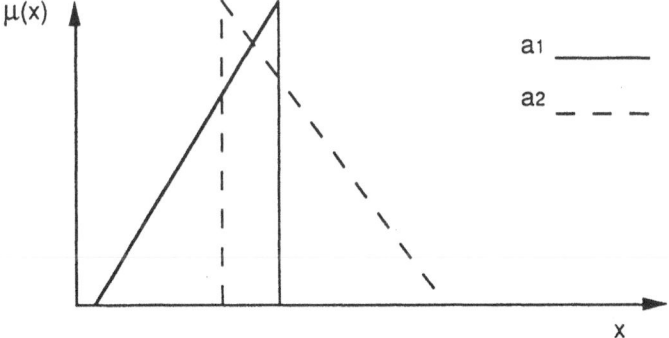

Figure 5.6 Counterintuitive Results
in Baas-Kwakernaak Model

A different approach based on linguistic solutions obtained by means of linguistic approximation techniques is described in Bonissone [1982] and Tong and Bonissone [1984]. This method can be divided into two steps:
• determination of fuzzy decision sets,

114

- truth qualification and linguistic approximation.

In the first step, indices, expressing the overall degree to which each of the alternatives dominates the other ones, are computed. One should note that these indices do not take into account the shape of the membership functions, but only one point is considered. For example, given the three fuzzy ratings of Figure 5.7, R_3 and R_2 would dominate R_1 to the same degree.

In the second step, a linguistic "label" is assigned to each action by means of linguistic approximation techniques. In particular, the semantic similarity between some preselected sentences and the unlabelled fuzzy sets representing the various alternatives, is evaluated by means of the Bhattacharya distance computed on some moments of the membership functions. Output of the model is a linguistic statement containing three pieces of information:

- an alternative,
- a linguistic evaluation of the intensity of preference,
- the degree of truth (τ) of the linguistic statement.

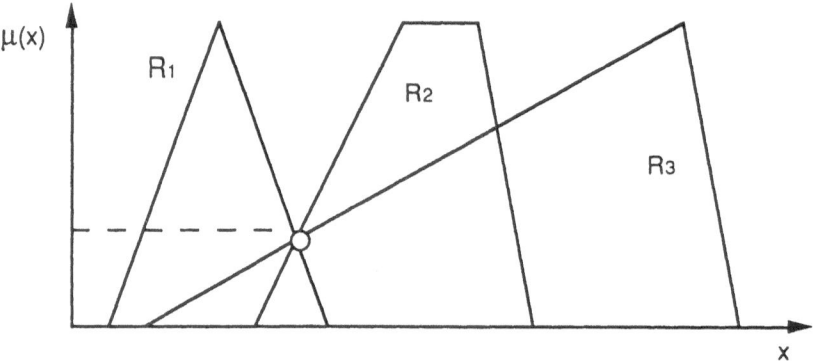

Figure 5.7   Dominance of Fuzzy Sets

Finally, we can synthesise the main characteristics of the above described methods as follows:

- most of them are limited to the use of triangular fuzzy numbers;
- the shape of the membership function is not taken into consideration or only a part of it is used (leading to a loss of information);
- a general problem is the one of the "sensitivity" (degree of discrimination) of the solutions[3];

---

[3] The degree of discrimination "refers to the capability of a method to differentiate between alternatives the ratings of which differ only slightly from each other [Zimmermann, 1986]".

- all these methods are utility based models (thus based on the complete transitive comparability axiom) aimed at finding a single "optimal" solution;
- most of these methods are limited to the use of only fuzzy information.

As one can see a key issue in multicriteria evaluation in a fuzzy environment is how to compare fuzzy sets. Chapter 6 deals with the problem of comparison of fuzzy sets; a new approach based on a semantic distance using areas instead of traditional intersections will be presented. In Chapter 7 a new fuzzy multicriteria method, based on some aspects of the partial comparability axiom, and on the new semantic distance will be presented. This method is an A.A.E. model where qualitative attributes are considered; in particular, the evaluations associated with each alternative can be real numbers, random variables with continuous and integrable density functions or fuzzy numbers (also with continuous and integrable membership functions).

CHAPTER 6

COMPARISON OF FUZZY SETS: A NEW SEMANTIC DISTANCE

6.1   The Problem of Comparison of Fuzzy Sets

In Chapter 5 it has been shown that an important problem in traditional discrete multicriteria methods in a fuzzy environment is the comparison of fuzzy sets. Here, first a concise overview of some other methods aiming at comparing fuzzy sets will be presented, and then a new approach based on the concept of a semantic distance, using areas instead of extreme values or values at the point of intersection, will be illustrated.

Most of the methods to rank fuzzy sets we survey here, have been developed in the field of fuzzy mathematical programming, where the main question to be answered is the comparison of the left and the right sides of a constraint (see section 4.6.1)

$$A_i(x) \leq B_i \qquad\qquad (6.1)$$

i.e. the comparison of two fuzzy numbers (generally flat fuzzy numbers).

A well-known concept widely used is the ρ-preference [Enta, 1982]:
let A_α and B_α be α-cuts of the fuzzy sets A and B, and ρ a credibility level with $0 \leq \rho \leq 1$. Then (see Figure 6.1)

$$A \leq_\rho B \text{ if } \quad \text{if sup } A_\alpha \leq \inf B_\alpha \ \forall \ \alpha \in [\rho, 1] \qquad\qquad (6.2)$$

The inequality relation can be considered a pessimistic index, since the possible surplus over B is neglected for all membership levels α with $\alpha < \rho$. It is possible that

$$\text{sup } A_\alpha < \text{sup} B_\alpha \quad \forall \ \alpha \in [0, 1] \qquad\qquad (6.3)$$

On the other hand it can happen that

$$\text{sup } A_\alpha > \text{sup} B_\alpha \quad \text{for some levels } \ \alpha < \rho \qquad\qquad (6.4)$$

118

Many similar approaches can be found in literature [De Campos & Moral, 1990; Gonzalez Munoz & Vila, 1990; Negoita et al., 1976; Ramik & Rimanek, 1985; Rommelfanger, 1989; Slowinski, 1986, 1990; Tanaka & Asai, 1984]. A good survey of most of these approaches can be found in Rommelfanger [1989a]. In particular, Slowinski tries to combine two indices: a pessimistic and an optimistic one, while Rommelfanger combines the pessimistic index of Slowinski with a fuzzy goal. One should note that all these approaches are characterised by the use of α-cuts and credibility levels.

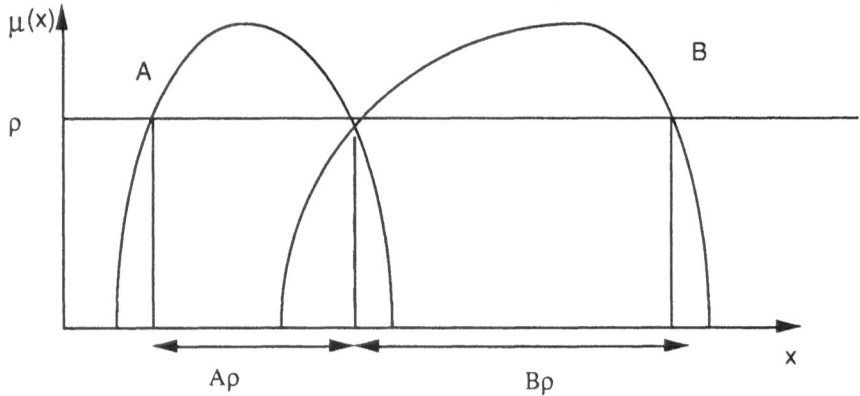

Figure 6.1   Inequality Relation $\leq \rho$

The use of areas for comparing fuzzy sets has been proposed by Li and Liu [1990] and Roubens [1990]. Li and Liu propose a new index called degree of coincidence to be used in order to compute the compatibility between two fuzzy sets in a different way than the one proposed by Zadeh. The degree of coincidence of a fuzzy set A with respect to a fuzzy set B is

$$w(A, B) = \frac{\int_U \min\{\mu_A(x)dx, \mu_B(x)dx\}}{\int_U \mu_B(x)dx} \qquad (6.5)$$

where U is the Universe of discourse in which the two fuzzy sets are defined. Note that in general this index is not symmetric.

Roubens computes the degree to which a fuzzy number is greater or equal to another fuzzy number, on the base of compensation of areas determined by the membership functions. However, one should note that if the case of L-R fuzzy numbers is considered, the comparison of areas is reduced to the comparison of upper and lower bounds of α-cuts. Let us consider a normalised convex fuzzy number A defined by the membership function $\mu_A(x)$, $x \in R$. The α-level sets $I(a, \alpha) = \{x \in R : \mu_A(x) \geq \alpha\}$ are convex subsets of R. The lower and upper limits of any α-level set $I(a, \alpha)$ are represented by inf $I(a, \alpha)$ and sup $I(a, \alpha)$, both limits are supposed to be finite. Roubens defines a transitive compensatory $\geq$ in the sense that value of α for which

$$\text{inf } I(a, \alpha) + \text{sup } I(a, \alpha) \geq \text{inf } I(b, \alpha) + \text{sup } I(b, \alpha) \tag{6.6}$$

are compensating those values for which the inequality is not satisfied.

Let A and B be two normalised convex fuzzy numbers, then

$$S_L(A \geq B) = \int_{U_1(a,b)} [\text{inf } I(a, \alpha) - \text{inf } I(b, \alpha)] \, d\alpha \tag{6.7}$$

where $U_1(a, b)$ being the subset of $[0, 1]$ such as

$$\{\alpha : \text{inf } I(a, \alpha) \geq \text{inf } I(b, \alpha)\} \tag{6.8}$$

$$S_R(A \geq B) = \int_{V_1(a,b)} [\text{sup } I(a, \alpha) - \text{sup } I(b, \alpha)] \, d\alpha \tag{6.9}$$

where $V_1(a, b)$ being the subset of $[0, 1]$ such as

$$\{\alpha : \text{sup } I(a, \alpha) \geq \text{sup } I(b, \alpha)\} \tag{6.10}$$

The degree to which $A \geq B$ is:

$$
\begin{aligned}
C(A \geq B) \quad &= S_L(A \geq B) + S_R(A \geq B) - S_L(B \geq A) - S_R(B \geq A), \quad \text{if positive} \\
&= 0 \text{ otherwise} \tag{6.11}
\end{aligned}
$$

Roubens proves that in the case of L-R fuzzy numbers, the comparison of areas is reduced to the comparison of limits of α-level sets, then a series of exact formulae can be obtained.

Finally, it should be noted that psychologists generally consider the expected value as a good representation of a fuzzy set [Zétényi, 1988].

The expected value of a fuzzy set A is equal to:

$$E(\mu_A(x)) = \frac{\int_{-\infty}^{+\infty} x\mu_A(x)dx}{\int_{-\infty}^{+\infty} \mu_A(x)dx} \tag{6.12}$$

provided that the integral converges absolutely; that is

$$\int_{-\infty}^{+\infty} |x\mu_A(x)| \, dx < \infty \tag{6.13}$$

otherwise, A has no finite expected value.

6.2 A New Semantic Distance

In general, a *semantic distance* S_d between two fuzzy sets, A and B, mirrors a possibility degree of equality between two fuzzy sets or a similarity degree between them. The larger the distance the smaller the possibility degree of equality. The most common distance is the so-called Hamming distance.
For the discrete case it is defined as:

$$S_d(A, B) = \sum_{i=1}^{n} |\mu_A(x_i) - \mu_B(x_i)| \tag{6.14}$$

and for the continuous case:

$$S_d(A, B) = \int_{-\infty}^{+\infty} |\mu_A(x) - \mu_B(x)| \, dx \tag{6.15}$$

For the continuous case, another possible approach is the computation of some moments of the membership distributions of the fuzzy sets, after which the similarity can be evaluated by means of traditional distances such as the Euclidean distance, the Bhattacharya distance, the Mahalanobis distance and so on [Dubois & Prade, 1980; Gupta & Sanchez, 1982]. Of course, in this case two problems have to be faced, viz. the correct selection of moments and the correct selection of the distance function.

Here we will illustrate a new semantic distance that is useful in the case of continuous, convex membership functions allowing also a definite integration. In order to compute such a distance, it is necessary that the area bounded by the membership function must be equal to 1. Generally, it is possible to change a membership function proportionally by multiplying it by a constant $k \in R^+$, with $k \leq 1$ for normal fuzzy sets and $k \leq 1/m_A$ for subnormal fuzzy sets ($m_A = \max_{x \in X} \mu_A(x)$) [Ragade & Gupta, 1977].

If $\mu_{A_1}(x)$ and $\mu_{A_2}(x)$ are two membership functions, we can write

$$f(x) = k_1 \mu_{A_1}(x) \quad \text{and} \tag{6.16}$$

$$g(y) = k_2 \mu_{A_2}(x) \tag{6.17}$$

where $f(x)$ and $g(y)$ are two functions obtained by rescaling the ordinates of $\mu_{A_1}(x)$ and $\mu_{A_2}(x)$ through k_1 and k_2, such as

$$\int_{-\infty}^{+\infty} f(x)\, dx = \int_{-\infty}^{+\infty} g(y)\, dy = 1 \tag{6.18}$$

The distance between all points of the membership functions is computed as follows:

$$\text{if } f(x) : X = [x_L, x_U] \quad \text{and} \quad g(y) : Y = [x_{L'}, x_{U'}] \tag{6.19}$$

(where of course sets X and Y can be non-bounded from one or either sides), then

$$S_d(f(x), g(y)) = \int_x \int_y |x - y|\, f(x)g(y)\, dy\,dx \tag{6.20}$$

6.2.1 Properties

In this section we will show the properties fulfilled by the semantic distance presented in (6.20). It is trivial to prove that this distance satisfies the properties of non-negativity and symmetry; the fulfilment of the property of triangle inequality can be proven as follows [Munda et al., 1993b].

Let us assume 3 functions:

$f(x) : X \rightarrow R^+, g(y) : Y \rightarrow R^+$ and $h(z) : Z \rightarrow R^+.$ (6.21)

we also assume that

$X \cap Y \cap Z \neq \emptyset$ (6.22)

We first prove that $\forall\ x \in X, \forall\ y \in Y$ and $\forall\ z \in Z$, the relationship

$|x-y| + |y-z| \geq |x-z|$ (6.23)

is always true.

The total number of possible cases is 3!

$x \geq y \geq z \rightarrow \quad (x-y) + (y-z) - (x-z)=0$ (6.24)

$x \geq z \geq y \rightarrow \quad (x-y) + (-y+z) - (x-z)= 2(z-y) \geq 0$ (6.25)

$y \geq x \geq z \rightarrow \quad (-x+y) + (y-z) - (x-z)= 2(y-x) \geq 0$ (6.26)

$y \geq z \geq x \rightarrow \quad (-x+y) + (y-z) - (-x+z)= 2(y-z) \geq 0$ (6.27)

$z \geq x \geq y \rightarrow \quad (x-y) + (-y+z) - (-x+z)= 2(x-y) \geq 0$ (6.28)

$z \geq y \geq x \rightarrow \quad (-x+y) + (-y+z) - (-x+z)=0$ (6.29)

therefore $|x-y| + |y-z| - |x-z| \geq 0\ \ \forall\ x \in X, \forall\ y \in Y$ and $\forall\ z \in Z$ (6.30)

Since $f(x) \geq 0, g(y) \geq 0$ and $h(z) \geq 0$ it follows:

$$\int_x \int_y \int_z\ \left[\, |x-y| + |y-z| - |x-z| \,\right] f(x)g(y)h(z)\ dzdydx \geq 0$$ (6.31)

This integral can be decomposed as follows:

$$\int_x \int_y \int_z \ |x-y| f(x)g(y)h(z) \ dzdydx \ + \int_x \int_y \int_z \ |y-z| f(x)g(y)h(z) \ dzdydx \ -$$

$$\int_x \int_y \int_z \ |x-z| f(x)g(y)h(z) \ dzdydx \qquad (6.32)$$

Since these triple integrals can be computed by means of iterated integrals and since, because of equation (6.18)

$$\int_x \ f(x) \ dx = \int_y \ g(y) \ dy = \int_z \ h(z) \ dz = 1 \qquad (6.33)$$

it follows that the sum of integrals in (6.32) is equal to:

$$\int_x \int_y \ |x-y| f(x)g(y) \ dydx \ + \int_y \int_z \ |y-z| g(y)h(z) \ dzdy \ -$$

$$\int_x \int_z \ |x-z| f(x)h(z) \ dzdx \qquad (6.34)$$

Therefore, we find the result:

$$S_d[f(x), g(y)] + S_d[g(y), h(z)] - S_d[f(x), h(z)] \geq 0 \qquad (6.35)$$

$$\text{or}$$

$$S_d[f(x), g(y)] + S_d[g(y), h(z)] \geq S_d[f(x), h(z)] \qquad \text{Q.E.D.} \qquad (6.36)$$

When both variables x and y are defined in the same interval, i.e. $x_L = x_{L'}$ and $x_U = x_{U'}$, for reasons of consistency, it is necessary to prove that the distance metric assumes a value different from zero. For simplicity, we write $x_L = x_{L'} = a$ and $x_U = x_{U'} = b$ (a<b) (see Figure 6.2). Let us consider the algebraic expression

$$|x-y| f(x) g(y) \qquad (6.37)$$

This product assumes the value zero if at least one of its three elements is zero. Concerning f(x) and g(y) we assume that:

124

$$\begin{cases} f(x)>0 & \forall\ x\in\ (a,b) \end{cases} \tag{6.38}$$

$$\begin{cases} f(x)=0 & \text{if } x=a \text{ or } x=b \qquad\text{and} \end{cases} \tag{6.39}$$

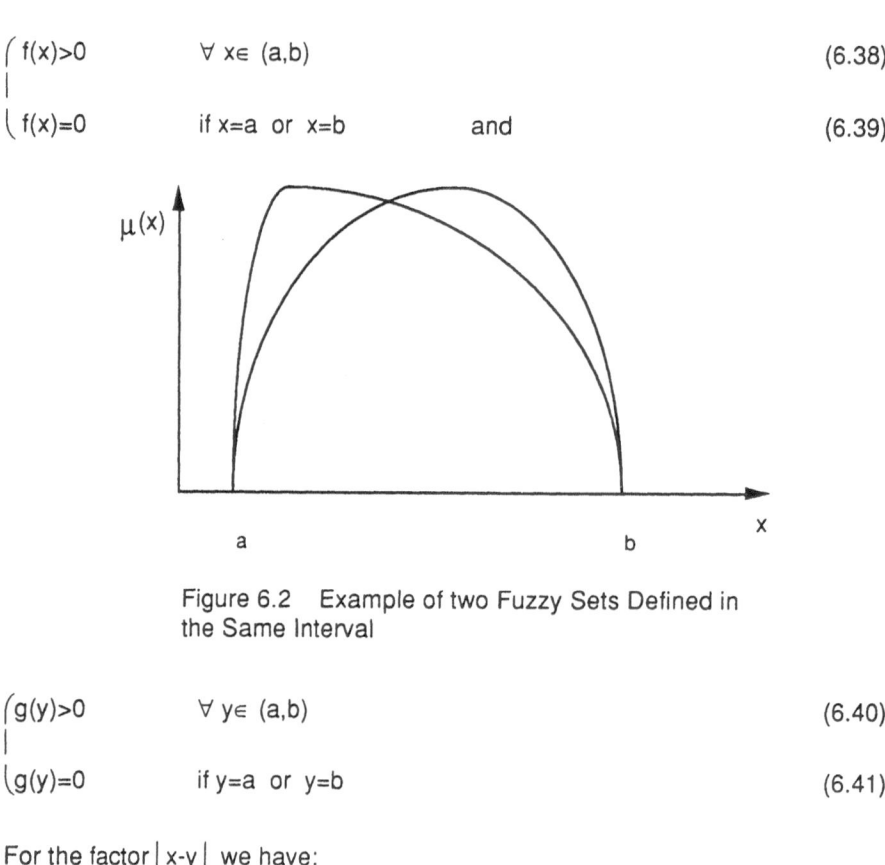

Figure 6.2   Example of two Fuzzy Sets Defined in the Same Interval

$$\begin{cases} g(y)>0 & \forall\ y\in\ (a,b) \end{cases} \tag{6.40}$$

$$\begin{cases} g(y)=0 & \text{if } y=a \text{ or } y=b \end{cases} \tag{6.41}$$

For the factor $|x-y|$ we have:

$$|x-y|>0 \qquad \forall\ x\neq y \in [a,b] \tag{6.42}$$

$$|x-y|=0 \qquad \text{if } x=y \tag{6.43}$$

On the basis of (6.38) - (6.43) it is possible to conclude that

$$|x-y|\ f(x)\ g(y)>0 \quad \forall\ x\neq y \in (a,b) \qquad\text{and} \tag{6.44}$$

$$|x-y|\ f(x)\ g(y)=0 \quad \text{elsewhere} \qquad\qquad \text{that is} \tag{6.45}$$

$$\forall\ x=y \in [a,b] \text{ or } x\neq y \text{ with } x=a \text{ or } x=b \text{ or } y=a \text{ or } y=b \tag{6.46}$$

If we take into consideration the summation of all

$$|x-y|\ f(x)\ g(y), \qquad \forall\ x\in [a,b] \text{ and } \forall\ y\in [a,b] \tag{6.47}$$

i.e. the integral $\int_x \int_y |x-y| f(x)g(y) \, dydx$, since it is not $|x-y| f(x) \, g(y)=0$

$\forall \, x \in [a,b]$ and $\forall \, y \in [a,b]$, it is

$$\int_x \int_y |x-y| f(x)g(y) \, dydx > 0 \qquad \text{Q.E.D.} \qquad\qquad (6.48)$$

It has to be noted that $S_d (f(x), g(y))=0$ if and only if $x=y \, \forall \, x \in [a,b]$ and $\forall \, y \in [a,b]$, that is if and only if

$$a=x=y=b \qquad\qquad (6.49)$$

i.e. if x and y are two equal crisp real numbers.

Finally, one has to note that if the distance between a fuzzy number and itself is computed, we impose the condition $S_d (f(x), f(x))=0$ by definition.

6.2.2 Results

As a special case, we consider first the case where the intersection of two membership functions is empty.

If $x>y \, \forall \, x \in X$ and $\forall \, y \in Y$, it follows that a continuous function in two variables is defined over a rectangle. Therefore, the double integral can be calculated as iterated single integrals:

$$\int_x \int_y |x-y| f(x)g(y) \, dydx= \qquad\qquad (6.50)$$

$$=\int_x \int_y (x-y)f(x)g(y) \, dydx= \qquad\qquad (6.51)$$

$$=\int_x \int_y [x \, f(x) \, g(y) - y \, f(x) \, g(y)] \, dydx= \qquad\qquad (6.52)$$

$$=\int_x x \, f(x) \, dx - \int_x f(x)E(y) \, dx = \qquad\qquad (6.53)$$

$$=E(x) - E(y)= \tag{6.54}$$

$$=| E(x)-E(y) | \tag{6.55}$$

where $E(x)$ and $E(y)$ are the expected values of the two membership functions; the latter result is true, since $x>y$.

Therefore, when the intersection is empty, their distance is equal to the distance between their expected values. When the intersection between two fuzzy sets is not empty, their distance, is greater than the difference between the expected values since $| x-y |$ is always greater than $(x-y)$. In this case one finds

$$S_d(f(x), g(y))=\int_{-\infty}^{\infty} \int_{x}^{\infty} (y-x) f(x)g(y) \, dydx + \int_{-\infty}^{\infty} \int_{-\infty}^{x} (x-y) f(x)g(y) \, dydx \tag{6.56}$$

This is the case of a double integral over a general region; since this is neither vertically simple nor horizontally simple, it is not possible the computation by means of iterated integration, but it is necessary to take the limit of the Rieman sum. This problem can be easily overcome by means of numerical analysis. In the next section, a Monte Carlo type numerical procedure for the computation of such a distance is illustrated.

This property of being dependent on the intersection is quite interesting since the overlapping area between two fuzzy sets is a key issue for determining how difficult their comparison is (when the intersection between two fuzzy sets is empty their comparison is unambiguous). In particular, since by means of the semantic distance presented here, when the intersection between two fuzzy sets is not empty, their distance is different from the difference between their expected values, a theoretical justification of the importance of the area of the intersection can be drawn.

In order to illustrate the results of this distance function, the following numerical application can be useful. Let us take into consideration the following two fuzzy sets:

$$\mu_A(x)=(1+((x-25)/5)^2)^{-1} \qquad\qquad x \in [25, 100] \tag{6.57}$$

and

$$\mu_B(x)= \begin{cases} 0 & \text{if } x \in [0, 50] \\ (1 + ((x - 50)/5)^{-2})^{-1} & \text{if } x \in (50, 100] \end{cases} \qquad (6.58)$$

Computing the expected values of the two fuzzy sets taken into consideration here, the following results are obtained:

$$E(\mu_A(x))=255.785/7.521=34.009 \qquad (6.59)$$

$$E(\mu_B(x))=3325/42.644=77.971 \qquad (6.60)$$

Therefore by means of their expected values the distance between the two fuzzy sets is

$$E(\mu_B(x)) - E(\mu_A(x))=43.962 \qquad (6.61)$$

Taking into consideration the semantic distance proposed here (computed by means of the numerical algorithm shown in section 6.3 with 1000 iterations), the following result is obtained:

$$S_d (\mu_A(x), \mu_B(x))= 44.254 \qquad (6.62)$$

As one can see, since the intersection between the two fuzzy sets is not empty, the result obtained by our distance function is different from the simple difference between their expected values.

From a theoretical point of view, the following conclusions can be drawn:

(1) the absolute value metric is a particular case of this type of distance;
(2) the expected value is obtained as a representation of fuzzy sets only when their intersection is empty;
(3) the comparison between a fuzzy number and a crisp number is equal to the difference between the expected value of the fuzzy number and the value of the crisp number considered;
(4) when the intersection between two fuzzy sets is not empty, their distance is greater than the difference between their expected values;
(5) in case of fuzzy information represented by L-R fuzzy numbers, when their intersection is empty, their distance is equal to the distance between their

middle values only when they are symmetric; otherwise, their expected values are obtained;

(6) by means of this semantic distance, the problem of the use of only one side of the membership functions, common to most of the traditional fuzzy multicriteria methods, is overcome;

(7) a theoretical justification of the importance of the area of the intersection in the comparison between two fuzzy sets can be drawn.

As one can see, the semantic distance proposed here overcomes various weak points inherent in traditional comparison methods. However, we also have to admit that the axiomatization is not complete; moreover the property of being dependent on the area of the intersection seems to conflict with the property that when the distance between a fuzzy number and itself is computed, this is equal to zero.

6.3 A Numerical Algorithm for the Computation of the Semantic Distance

In this section, in order to compute the semantic distance proposed, a Monte Carlo type of numerical algorithm will be described.

<u>Assumptions:</u>

Our starting point is:

$$f(x) : X = [x_L, x_U] \rightarrow M \tag{6.63}$$

$$g(y) : Y = [x_{L'}, x_{U'}] \rightarrow M \tag{6.64}$$

where M is the membership space.

All $x \in X$ and all $y \in Y$ can be obtained by means of a random generator that supplies uniformly distributed numbers $r \in [0, 1]$ and then

$$x = rx_L + (1-r)x_U \qquad \text{and} \tag{6.65}$$

$$y = rx_{L'} + (1-r)x_{U'} \tag{6.66}$$

The probability to obtain a point P inside e.g. $f(x)$, whose value on the x-axis is x_0, depends on the shape of the function; therefore, an auxiliary variable Z is introduced, with values $z \in [0, \text{max}f(x)]$, by a random generator.

<u>Procedure:</u>

STEP 1: draw a random number r_0 ($0 \le r_0 \le 1$);

STEP 2: $x_0 = r_0 x_L + (1-r_0)x_U$;

STEP 3: draw a random number z_0 ($0 \le z_0 \le \text{max}f(x)$);

STEP 4: if: $z_0 \le f(x_0)$ then go to next step,

                $z_0 > f(x_0)$ then return to step 1;

STEP 5: draw a random number r_1 ($0 \le r_1 \le 1$);

STEP 6: $y_1 = r_1 x_{L'} + (1-r_1)x_{U'}$;

STEP 7: draw a random number z_1 ($0 \le z_0 \le \text{max}g(y)$);

STEP 8: if: $z_1 \le g(y_1)$ then compute $|x_0 - y_1|$;

                $z_1 > g(y_1)$ then return to step 5;

This procedure must be repeated many times. If n values of $|x_i - y_i|$ are obtained, then the distance between two fuzzy sets is approximately equal to the arithmetic mean of all the points bounded by their respective membership functions obtained by drawing random numbers.

In more formal terms, it is

$$\int_x \int_y |x - y| f(x)g(y) \, dy \, dx \cong \frac{\sum_{i=1}^{n} |x_i - y_i|}{n} \tag{6.67}$$

with i= 1, 2, ..., n.

6.4   A Generalisation of the Minkowski p-metric

It is interesting to note that also the *stochastic information* represented by means of density functions can be taken into account by means of this distance function. Of course in this case the condition

$$\int_x f(x)\,dx = 1 \tag{6.68}$$

is always true.

Generally, for crisp cardinal evaluations, the Minkowski p-metric is considered: given any two points x, y $\in R^N$, their distance is given by:

$$|\bar{x}-\bar{y}| = \left[\sum_{n=1}^{N}|x_n - y_n|^p\right]^{\frac{1}{p}} \tag{6.69}$$

$p\in\{1, 2, \dots\}\cup\{\infty\}$

It is then clear that we have:
for p=1 an absolute value metric (completely compensatory);
for p=2 an Euclidean metric (partially compensatory);
for p→∞ the Tchebycheff metric (completely non compensatory).

For the problem of mixed information, we propose the following generalisation of the Minkowski metric: given R crisp evaluations and I stochastic and/or fuzzy evaluations of the performance of each alternative, the distance between two elements x and y is:

$$|\bar{x}-\bar{y}| = \left[\sum_{r=1}^{R}|x_r - y_r|^p + \sum_{i=1}^{I}\left(\iint |x - y| f_i(x) g_i(y)\,dydx\right)^p\right]^{\frac{1}{p}} \tag{6.70}$$

(of course, first all the evaluations must be normalised).

In our opinion, this distance has the advantage of dealing simultaneously with different kinds of information, so *it can be an appropriate tool in order to increase the equivalence of the procedures used for the different types of available information.*

CHAPTER 7

MULTICRITERIA EVALUATION IN A FUZZY ENVIRONMENT: THE *NAIADE* METHOD

7.1   Introduction

Since the complexity of environmental problems is high, there is a clear need for models offering a comprehensible and operational representation of a real-world environmental system. Qualitative aspects are hard to deal with in traditional models and therefore there is a clear need for methods that are able to take into account information of a "mixed" type (both qualitative and quantitative measurements). Traditional qualitative multicriteria approaches take into consideration the case where information on an ordinal scale is present. A problem, related to all multicriteria methods that try to take mixed information into account is the problem of equivalence of the procedures used in standardising the various evaluations of the performance of alternatives according to different criteria. Another problem related to the available information concerns the uncertainty (stochastic and/or fuzzy) contained in this information. Therefore, the combination of different levels of measurement with different types of uncertainty has to be considered as an important research issue in multicriteria evaluation.

Our overview of multicriteria evaluation methods in a fuzzy environment, shows that traditional fuzzy multicriteria methods are utility based models; most of these methods are limited to the use of only fuzzy information (often only triangular fuzzy numbers) and a key issue is how to compare fuzzy sets.

Given a "consistent family" of evaluation criteria $G=\{g_m\}$, m=1,2,..., M, and a finite set $A=\{a_n\}$, n=1, 2,..., N of potential alternatives (actions)[1], here we will present a new multicriteria method based on some aspects of the partial comparability axiom, called NAIADE (**N**ovel **A**pproach to **I**mprecise **A**ssessment and **D**ecision **E**nvironments). It is a discrete multicriteria method whose impact (or evaluation) matrix may include either crisp, stochastic or fuzzy measurements of the performance of an alternative a_n with respect to a judgement criterion g_m. No traditional weighting of criteria is used in this method. In developing such a fuzzy multicriteria method, throughout this Chapter the assumption is made that a

---

[1] In the terminology introduced by Vansnick [1990], the decision model considered can be defined *"model A.A.E."* (Alternatives, Attributes, Evaluators), where qualitative attributes are considered. In particular, here, the evaluations associated with each alternative can be real numbers, random variables with continuous and integrable density functions or fuzzy numbers (also with continuous and integrable membership functions).

higher value of a criterion is preferred to a lower one (the higher, the better). From an empirical point of view, this model is particularly suitable for economic-ecological modelling incorporating various degrees of precision of the variables taken into consideration. From a methodological point of view, two main issues will be faced here:

- the problem of equivalence of the used procedures in order to standardise the various evaluations (of a mixed type) of the performance of alternatives according to different criteria;
- the problem of comparison of fuzzy numbers typical of all fuzzy multicriteria methods.

The whole procedure can be divided in three main steps [Munda et al., 1995a]:

(1)   pairwise comparison of alternatives,
(2)   aggregation of all criteria,
(3)   evaluation of alternatives.

We will start the presentation by introducing the concept of a "fuzzy preference relation".

7.2   The Notion of a Fuzzy Preference Relation

If A is assumed to be a finite set of N alternatives, a *fuzzy preference relation* is an element of the NxN matrix R= (r_{ij}), i.e.

$$r_{ij}= \mu_R (a_i, a_j) \quad \text{with } i,j= 1,2, ..., N \quad \text{and } 0 \leq r_{ij} \leq 1 \qquad (7.1)$$

$r_{ij}= 1$ indicates the maximum degree of preference of a_i over a_j; each value of r_{ij} in the open interval (0.5, 1) indicates a definite preference of a_i to a_j (a higher value means a stronger intensity); $r_{ij}= 0.5$ indicates the indifference between a_i and a_j. This definition implies that fuzzy preference relations can be used as mathematical models of preference and *intensity of preference*.

Usually, fuzzy preference relations are assumed to satisfy two properties:

(1)   reciprocity, i.e. $r_{ij}+r_{ji}=1$;
(2)   max-min transitivity, i.e. if a_i is preferred to a_j and a_j is preferred to a_k, then a_i should be preferred to a_k with at least the same intensity, that is

$$r_{ij} \geq 0.5, \quad r_{jk} \geq 0.5 \Rightarrow r_{ik} \geq \min(r_{ij}, r_{jk}) \qquad (7.2)$$

Some authors maintain that in order to be consistent, fuzzy preference relations might have to satisfy other types of transitivity as: restricted max-max transitivity, additive transitivity, and multiplicative transitivity [Tanino, 1988]. It should be noted that in these cases a close relationship between cardinal utility theory and fuzzy relations exist.

7.3 Pairwise Comparison of Alternatives

The comparison between the criterion scores of each pair of actions is carried out by means of the semantic distance described in (6.20). As shown in Chapter 6, this distance is an efficient approach for both the comparison of fuzzy numbers and to increase the equivalence of the procedures used for different types of available information.

In section 4.7.2.1, it has been noted that the modelling procedure based on the notion of a pseudo-criterion, may present a serious lack of stability. Such undesirable discontinuities make a sensitivity analysis (or robustness analysis) necessary; however, this important analysis step is quite complex to manage because of the combinatorial nature of the various sets of data. One should combine variations of two thresholds and k scores of the M criteria. The use of fuzzy set approaches, since small variations of input data (scores, thresholds) will modify in a continuous way the resulting preference model, can allow one to avoid these drawbacks [Perny & Roy, 1992]. A first possible approach is to associate a fuzzy outranking relation with a pseudo-criterion as in ELECTRE III (see Figure 7.1), or to use a generalised criterion as in PROMETHEE. It has to be noted that both methods require that some parameters have to be fixed. A second approach may be the use of fuzzy criterion scores; however, in this case the problem of comparison of fuzzy numbers arises.

We propose a method based on both the underlying philosophy of a "fuzzy pseudo-criterion" and the use of fuzzy criterion scores. In section 5.9, we have seen that by using fuzzy relations a formal analysis of imprecise linguistic relations is allowed. Here, fuzzy binary relations will be used to model different possible preference/indifference situations.

Six different fuzzy relations are simultaneously considered:

1) much greater than ($\gg$)
2) greater than ($>$)
3) approximately equal to ($\cong$)
4) very equal to ($=$)

134

5) less than (<)
6) much less than (<<).

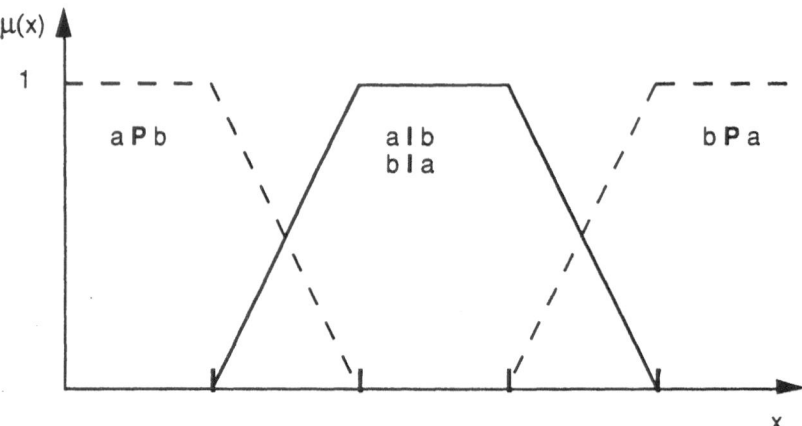

Figure 7.1   Fuzzy Relations with Thresholds and Crisp Scores

Analytically, membership functions pertaining to the 6 fuzzy relations can be formulated in the equations below (of course if one wishes, the shape of the membership functions of these fuzzy relations may be defined in different ways too). In order to improve the understanding a graphical illustration is presented in Appendix 1. According to fuzzy philosophy, no precise boundary is established, thus the decision-maker is not asked to evaluate thresholds, which is always a difficult and perhaps arbitrary process, although it should be admitted that the choice of the membership functions is also somewhat arbitrary.

Given a criterion g_m and a pair of alternatives a, b, it is for crisp and fuzzy criterion scores respectively:

a) <u>crisp criterion scores</u>

1) greater than

$$\mu_>(a,b)_m = \begin{cases} 0 & \text{if } x-y \leq 0 \\ [1+c_1(x-y)^{-2}]^{-1} & \text{if } x-y > 0 \end{cases} \qquad (c_1 \in R^+) \quad (7.3)$$

2) much greater than

$$\mu_{>>}(a,b)_m= \begin{cases} 0 & \text{if } x-y\leq0 \\ [1+c_2(x-y)^{-2}]^{-2} & \text{if } x-y>0 \end{cases} \qquad (c_2\in R^+) \quad (7.4)$$

3) approximately equal to

$$\mu_{\cong}(a,b)_m= e^{-c_3(|x-y|)} \qquad (c_3\in R^+) \quad (7.5)$$

4) very equal to

$$\mu_{=}(a,b)_m= e^{-c_4(x-y)^2} \qquad (c_4\in R^+) \quad (7.6)$$

5) less than

$$\mu_{<}(a,b)_m= \begin{cases} [1+c_5(y-x)^{-2}]^{-1} & \text{if } x-y<0 \\ 0 & \text{if } x-y\geq0 \end{cases} \qquad (c_5\in R^+) \quad (7.7)$$

6) much less than

$$\mu_{<<}(a,b)_m= \begin{cases} [1+c_6(y-x)^{-2}]^{-2} & \text{if } x-y<0 \\ 0 & \text{if } x-y\geq0 \end{cases} \qquad (c_6\in R^+) \quad (7.8)$$

where $x=g_m(a)$ and $y=g_m(b)$.

b) <u>fuzzy and stochastic criterion scores</u>

1) greater than

$$\mu_{>}(a,b)_m= \begin{cases} 0 & \text{if } \iint(x-y)f(x)g(y)dydx\leq0 \\ \left[1+c_1\left(\iint(x-y)f(x)g(y)dydx\right)^{-2}\right]^{-1} & \text{if } \iint(x-y)f(x)g(y)dydx>0 \end{cases} \qquad (c_1\in R^+) \quad (7.9)$$

2) much greater than

$$\mu_{>>}(a,b)_m = \begin{cases} 0 & \text{if } \iint (x-y)f(x)g(y)dydx \leq 0 \\ \\ \left[1+c_2\left(\iint (x-y)f(x)g(y)dydx\right)^{-2}\right]^{-2} & \text{if } \iint (x-y)f(x)g(y)dydx > 0 \end{cases} \qquad (c_2 \in R^+) \quad (7.10)$$

3) approximately equal to

$$\mu_{\cong}(a,b)_m = e^{-c_3\left(\iint |x-y|\, f(x)g(y)dydx\right)} \qquad (c_3 \in R^+) \quad (7.11)$$

4) very equal to

$$\mu_{=}(a,b)_m = e^{-c_4\left(\iint |x-y|\, f(x)g(y)dydx\right)^2} \qquad (c_4 \in R^+) \quad (7.12)$$

5) less than

$$\mu_{<}(a,b)_m = \begin{cases} \left[1+c_5\left(\iint (y-x)f(x)g(y)dxdy\right)^{-2}\right]^{-1} & \text{if } \iint (x-y)f(x)g(y)dydx < 0 \\ \\ 0 & \text{if } \iint (x-y)f(x)g(y)dydx \geq 0 \end{cases} \qquad (c_5 \in R^+) \quad (7.13)$$

6) much less than

$$\mu_{<<}(a,b)_m = \begin{cases} \left[1+c_6\left(\iint (y-x)f(x)g(y)dxdy\right)^{-2}\right]^{-2} & \text{if } \iint (x-y)f(x)g(y)dydx < 0 \\ \\ 0 & \text{if } \iint (x-y)f(x)g(y)dydx \geq 0 \end{cases} \qquad (c_6 \in R^+) \quad (7.14)$$

where $f(x) = g_m(a)$ and $g(y) = g_m(b)$.

One should note that all the fuzzy relations are defined in a perfect equivalent way for both crisp and fuzzy (stochastic) evaluations; thus the semantic distance used seems a promising tool to arrive at equivalence in the treatment of different kinds of criterion scores.

The relations $\mu_{>}(a,b)_m$, $\mu_{>>}(a,b)_m$, $\mu_{<}(a,b)_m$ and $\mu_{<<}(a,b)_m$ are derived from values satisfying the condition of additive transitivity for both crisp and stochastic criterion scores. Thus it is trivial to prove that all these relations are max-min transitive, however the property of reciprocity does not hold, thus these are not fuzzy preference relations in a "strictu sensu". On the contrary, the relations

$\mu_{\leqq}(a,b)_m$ and $\mu_{=}(a,b)_m$ are simply *resemblance relations*, which are reflexive and symmetric but no transitivity is implied.

Finally, it has to be noted that by the construction of these 6 relations, also the normalisation (i.e. elimination of the scaling effects) of the criterion scores is achieved by defining the parameter c in interaction with the decision-maker (e.g. by defining the crossover point[2]).

7.4 Aggregation of All Criteria

Given the above information on the pairwise performance of the alternatives according to each single criterion, it is necessary to aggregate these evaluations in order to take all criteria into account simultaneously. For this purpose, some results obtained in the field of group decision theory can be useful [Tanino, 1988].

The simplest way of aggregating such individual values is the following:

$$\mu_{\cdot}(a,b)= \frac{1}{M}\sum_{m=1}^{M}\mu_{\cdot}(a,b)_m \tag{7.15}$$

where $\mu_{*}(a,b)_m$ indicates the evaluation of a given fuzzy relation for each pair of actions according to the m-th criterion. A disadvantage of this approach is that the diversity among the assessments of single fuzzy relations is not considered (since the preference intensities completely compensate one another).

Another approach can be found in the use of an α-level nonfuzzy preference relation; this is defined by

$$\mu_{\cdot}(a, b)_m(\alpha)= \begin{cases} 1 & \text{if } \mu_{\cdot}(a, b)_m \geq \alpha \\ 0 & \text{if } \mu_{\cdot}(a, b)_m < \alpha \end{cases} \tag{7.16}$$

However, the application of this rule in our problem would imply the loss of any information of the $\mu_{*}(a,b)_m$ on the intensity of preference.

Therefore, we propose to use the following equation:

$$\mu_{\cdot}(a,b)= \frac{\sum_{m=1}^{M}\max\left(\mu_{\cdot}(a,b)_m - \alpha, 0\right)}{\sum_{m=1}^{M}\left|\mu_{\cdot}(a,b)_m - \alpha\right|} \tag{7.17}$$

---

[2] The crossover point indicates the value of the variable at which the membership function is equal to 0.5.

where α is a "minimum requirement" imposed on each fuzzy relation.

It is:

$0 \leq \mu_*(a,b) \leq 1$,   with

$\mu_*(a,b)=0$ if no $\mu_*(a,b)_m$ is greater than α;

$\mu_*(a,b)=1$ if $\mu_*(a,b)_m \geq \alpha \ \forall \ m$, and $\mu_*(a,b)_m > \alpha$ for at least one m.

The equation described in (7.17) has been used by Nakayama and others [1979] as the base of the so called "extended contributive rule method". They consider only the case $\alpha=0.5$ and individual additively transitive preference relations are assumed. These authors propose to evaluate the diversity of preferences of the various actors by means of measures of fuzziness (see Section 5.9). Bezdek and others [1978, 1979] also propose to use measures of fuzziness for fuzzy binary relations in a context of group decision theory; they use two measures called "average fuzziness" and "average certainty".

Entropy measures of fuzziness can be easily extended to the notion of fuzzy relations as shown in Gupta et al. [1977]. Thus to have information on the diversity among the assessments of the single fuzzy relations, according to each criterion, the entropy concept is useful; here, it is (see equation (7.20))

$$\mu_A(x) = \begin{cases} 0 & \text{if } \mu_*(a, b)_m - \alpha \leq 0 \\ \mu_*(a, b)_m & \text{if } \mu_*(a, b)_m - \alpha > 0 \end{cases} \tag{7.18}$$

where $\mu_A(x)$ is the membership function of a fuzzy set A on a Universe X, and the entropy is given by

$$H(A) = \frac{1}{N} \sum_{x \in X} \ln(x) \tag{7.19}$$

where N is the number of elements in X (X being finite) and ln(x) is the *incertitude* of the evaluation along scale x given by

$$\ln(x) = -[\mu_A(x) \log_2 \mu_A(x) + (1-\mu_A(x)) \log_2 (1-\mu_A(x))] \tag{7.20}$$

The entropy of a set A is 1 if for every x, $\mu_A(x)=0.5$ and is 0 if for every x, $\mu_A(x)=1$ or 0.

Entropy measures have been first proposed in information theory [Shannon, 1948]; applications in economics can be found in Theil [1967]. Entropy measures of fuzziness have been proposed by De Luca & Termini [1972] and Capocelli & De Luca [1973].

Now a kind of aggregate "fuzzy preference relation" can be obtained:

$$
\begin{pmatrix}
\mu_{>>}(a,b) & H(>>) \\
\mu_{>}(a,b) & H(>) \\
\mu_{\cong}(a,b) & H(\cong) \\
\mu_{=}(a,b) & H(=) \\
\mu_{<}(a,b) & H(<) \\
\mu_{<<}(a,b) & H(<<)
\end{pmatrix}
$$

where $\mu_*(a,b)$ is the overall evaluation of a given fuzzy relation for each pair of actions obtained by means of equation (7.17) and $H(*)$ is the associated entropy level. How to use such information in order to evaluate the different alternatives will be the topic of the next section.

7.5   Evaluation of the Alternatives

The information provided by a "fuzzy preference relation" can be used in different ways, e.g., the degree of truth (τ) of statements as: "according to most of the criteria

- **a** is better than **b**,
- **a** and **b** are indifferent,
- **a** is worse than **b**"

can be computed as follows.

The proportional linguistic quantifier "most" can be defined by the membership function (arbitrarily defined)

$$
\mu_{most}(\omega)=
\begin{cases}
1 & \text{if } \omega \geq 0.8 \\
3.33\omega - 1.66 & \text{if } 0.5 < \omega < 0.8 \\
0 & \text{if } \omega \leq 0.5
\end{cases}
\tag{7.21}
$$

It is evident that $\forall \omega \in [0, 1]$ if $\omega' > \omega'' \Rightarrow \mu_{most}(\omega') \geq \mu_{most}(\omega'')$; in other words it is a nondecreasing fuzzy quantifier [Fedrizzi & Kacprzyk, 1988; Zadeh, 1983].

The value ω is a function of the aggregate membership degree and its entropy level. Then after the transformation $C(*)=1-H(*)^3$, it is possible to define:

$$\omega(\mathbf{a}\text{ is better than }\mathbf{b})= \frac{\mu_{>>}(a,b)\wedge C(>>)+\mu_>(a,b)\wedge C(>)}{C(>>)+C(>)} \tag{7.22}$$

$$\omega(\mathbf{a}\text{ is worse than }\mathbf{b})= \frac{\mu_{<<}(a,b)\wedge C(<<)+\mu_<(a,b)\wedge C(<)}{C(<<)+C(<)} \tag{7.23}$$

$$\omega(\mathbf{a}\text{ and }\mathbf{b}\text{ are indifferent})= \frac{\mu_=(a,b)\wedge C(=)+\mu_\cong(a,b)\wedge C(\cong)}{C(=)+C(\cong)} \tag{7.24}$$

(where "$\wedge$" may be replaced by any other operator).

Thus concerning each pair of actions a global linguistic evaluation characterised by its degree of truth[4], is obtained.

If only a small number of alternatives is taken into account, these pairwise evaluations can be used directly by the decision-maker(s) in order to isolate a set of satisfactory solutions, or if in a given decisional environment there is a need to perform further elaboration in order to get a ranking of the alternatives, they offer useful complementary information in order to arrive at a final decision. If a ranking of the alternatives in a complete or partial preorder is needed (γ problem formulation), the basic idea of positive (leaving) and negative (entering) flows of the PROMETHEE methods [Brans et al., 1986] can be adapted to the features of our procedure. An axiomatic characterisation of the "leaving and entering flows method" can be found in Bouyssou and Perny [1990].

For each action we define:

$$\phi^+(a)=\frac{\sum_{n=1}^{N-1}\delta_n}{\sum_{n=1}^{N-1}C_n(>>)+\sum_{n=1}^{N-1}C_n(>)} \tag{7.25}$$

---

[3] This transformation is necessary in order to apply some computational rules of approximate reasoning as defined by Zadeh [1983].

[4] The use of truth-values is one of the possible ways (together with the combination with other membership functions or the application of a modifier) in which a membership function can be modified; this kind of transformation is defined by the *rule of truth-functional modification*. This rule may be stated in symbols as follows: $(x\text{ is }P)$ is $\tau \Rightarrow x$ is $\mu_P^{-1}\circ\tau$, where $\mu_P^{-1}\circ\tau$ is the composition of the nonfuzzy relation μ_P^{-1} with the unary fuzzy relation τ.

where $\delta_n=\mu_{>>}(a,x)\wedge C(>>) + \mu_>(a,x)\wedge C(>)$ (7.26)

and

$$\phi^-(a)=\frac{\sum\limits_{n=1}^{N-1}\Psi_n}{\sum\limits_{n=1}^{N-1}C_n(<<)+\sum\limits_{n=1}^{N-1}C_n(<)}$$ (7.27)

where $\psi_n=\mu_{<<}(a,x)\wedge C(<<) + \mu_<(a,x)\wedge C(<)$ (7.28)

Finally, it has to be noted that the rankings obtained by means of these two indices include a higher degree of discrimination than the pairwise linguistic evaluations, because additional information is used, viz. the relations of each alternative with all other ones, and no "minimum requirement threshold" ($\omega\leq 0.5$) like in the linguistic quantifiers is introduced.

7.6 An Illustrative Example

Let us take into consideration the transportation problem described in section 4.7.1.3 again. Now the qualitative information will be represented by means of fuzzy sets. The impact (or evaluation) matrix related to the above problem is supposed to be the following:

Criteria	Units	Alternatives		
		Highway (a_1)	Road/bus (a_2)	Train (a_3)
Costs	mln gld	200	250	400
Travel Time	linguistic	excellent	good	moderate
Capacity	mln km/year	20	30	40
NO$_x$ Emissions	ton/year	1000	750	100
Landscape	linguistic	bad	bad	moderate

Table 7.1   Fuzzy Evaluation Matrix of a Transportation Problem

In Appendix 2 the membership functions of the used linguistic variables are illustrated.

By applying NAIADE the following pairwise fuzzy relations are obtained:

COSTS

$\mu_{>>}(a_1, a_2)=0.23$	$\mu_{>>}(a_1, a_3)=0.87$	$\mu_{>>}(a_2, a_3)=0.80$
$\mu_{>}(a_1, a_2)=0.60$	$\mu_{>}(a_1, a_3)=0.96$	$\mu_{>}(a_2, a_3)=0.93$
$\mu_{\cong}(a_1, a_2)=0.17$	$\mu_{\cong}(a_1, a_3)\cong0$	$\mu_{\cong}(a_2, a_3)\cong0$
$\mu_{=}(a_1, a_2)\cong0$	$\mu_{=}(a_1, a_3)\cong0$	$\mu_{=}(a_2, a_3)\cong0$

TRAVEL TIME

$\mu_{>>}(a_1, a_2)=0.52$	$\mu_{>>}(a_1, a_3)=0.79$	$\mu_{>>}(a_2, a_3)=0.33$
$\mu_{>}(a_1, a_2)=0.60$	$\mu_{>}(a_1, a_3)=0.86$	$\mu_{>}(a_2, a_3)=0.50$
$\mu_{\cong}(a_1, a_2)=0.23$	$\mu_{\cong}(a_1, a_3)\cong0$	$\mu_{\cong}(a_2, a_3)=0.34$
$\mu_{=}(a_1, a_2)\cong0$	$\mu_{=}(a_1, a_3)\cong0$	$\mu_{=}(a_2, a_3)\cong0$

CAPACITY

$\mu_{\cong}(a_1, a_2)=0.09$	$\mu_{\cong}(a_1, a_3)\cong0$	$\mu_{\cong}(a_2, a_3)=0.09$
$\mu_{=}(a_1, a_2)\cong0$	$\mu_{=}(a_1, a_3)\cong0$	$\mu_{=}(a_2, a_3)\cong0$
$\mu_{<}(a_1, a_2)=0.80$	$\mu_{<}(a_1, a_3)=0.94$	$\mu_{<}(a_2, a_3)=0.8$
$\mu_{<<}(a_1, a_2)=0.62$	$\mu_{<<}(a_1, a_3)=0.87$	$\mu_{<<}(a_2, a_3)=0.62$

NO$_X$

$\mu_{\cong}(a_1, a_2)\cong0$	$\mu_{\cong}(a_1, a_3)\cong0$	$\mu_{\cong}(a_2, a_3)\cong0$
$\mu_{=}(a_1, a_2)\cong0$	$\mu_{=}(a_1, a_3)\cong0$	$\mu_{=}(a_2, a_3)\cong0$
$\mu_{<}(a_1, a_2)=0.94$	$\mu_{<}(a_1, a_3)\cong1$	$\mu_{<}(a_2, a_3)\cong1$
$\mu_{<<}(a_1, a_2)=0.90$	$\mu_{<<}(a_1, a_3)\cong1$	$\mu_{<<}(a_2, a_3)=0.98$

LANDSCAPE

$\mu_{\cong}(a_1, a_2)=1$	$\mu_{\cong}(a_1, a_3)=0.34$	$\mu_{\cong}(a_2, a_3)=0.34$
$\mu_{=}(a_1, a_2)=1$	$\mu_{=}(a_1, a_3)\cong0$	$\mu_{=}(a_2, a_3)\cong0$
$\mu_{<}(a_1, a_2)=0$	$\mu_{<}(a_1, a_3)=0.5$	$\mu_{<}(a_2, a_3)=0.5$
$\mu_{<<}(a_1, a_2)=0$	$\mu_{<<}(a_1, a_3)=0.33$	$\mu_{<<}(a_2, a_3)=0.33$

On the basis of these for each pair of actions the following aggregate values are obtained (see equations (7.17) and (7.20)):

$$\left(\begin{array}{ll} \mu_>(a_1, a_2)=0.40 & H(>)=0.38 \\ \mu_{>>}(a_1, a_2)=0.18 & H(>>)=0.20 \\ \mu_\cong(a_1, a_2)=0.49 & H(\cong)=0 \\ \mu_=(a_1, a_2)=0.36 & H(=)=0 \\ \mu_<(a_1, a_2)=0.55 & H(<)=0.17 \\ \mu_{<<}(a_1, a_2)=0.50 & H(<<)=0.28 \end{array}\right.$$

$$\left(\begin{array}{ll} \mu_>(a_1, a_3)=0.57 & H(>)=0.16 \\ \mu_{>>}(a_1, a_3)=0.54 & H(>>)=0.25 \\ \mu_\cong(a_1, a_3)=0.03 & H(\cong)=0.18 \\ \mu_=(a_1, a_3)=0 & H(=)=0 \\ \mu_<(a_1, a_3)=0.71 & H(<)=0.26 \\ \mu_{<<}(a_1, a_3)=0.68 & H(<<)=0.29 \end{array}\right.$$

$$\left(\begin{array}{ll} \mu_>(a_2, a_3)=0.47 & H(>)=0.27 \\ \mu_{>>}(a_2, a_3)=0.37 & H(>>)=0.32 \\ \mu_\cong(a_2, a_3)=0.08 & H(\cong)=0.36 \\ \mu_=(a_2, a_3)=0 & H(=)=0 \\ \mu_<(a_2, a_3)=0.70 & H(<)=0.34 \\ \mu_{<<}(a_2, a_3)=0.63 & H(<<)=0.40 \end{array}\right.$$

For each pair of actions, the following degrees of truth of a linguistic evaluation are obtained (see equations from (7.21) to (7.24):

a_1 is better than a_2	$\tau=0$
a_1 and a_2 are indifferent	$\tau=0$
a_1 is worse than a_2	$\tau=0.57$
a_1 is better than a_3	$\tau=0.67$
a_1 and a_3 are indifferent	$\tau=0$
a_1 is worse than a_3	$\tau=1$
a_2 is better than a_3	$\tau=0.53$

a_2 and a_3 are indifferent	$\tau=0$
a_2 is worse than a_3	$\tau=1$

These results are mainly due to four factors:
- the number of criteria in favour of an action;
- the degree of compensation allowed in the aggregation process (in this application we have used a minimum requirement for each criterion, equal to $\alpha=0.30$);
- the definition of the membership function of the linguistic operators;
- the aggregation operator chosen for the approximate reasoning operations (in this application we have used the "min" operator which is known as a representation of the logic "and", and therefore it is completely non-interactive (since a high value cannot compensate a low one)).

By computing the ϕ^+ and ϕ^- indices (see equations from (7.25) to (7.28)), the following results are obtained:

$\phi^+(a_1)=0.564$
$\phi^-(a_1)=0.817$

$\phi^+(a_2)=0.643$
$\phi^-(a_2)=0.680$

$\phi^+(a_3)=0.980$
$\phi^-(a_3)=0.656$

Then according to ϕ^+ the following ranking is obtained
$a_3 \rightarrow a_2 \rightarrow a_1$

and according to ϕ^- the following ranking is obtained
$a_3 \rightarrow a_2 \rightarrow a_1$

Therefore, the intersection gives as a final result the complete preorder

$$a_3 \rightarrow a_2 \rightarrow a_1$$

This ranking is a function of all actions taken into consideration; while on the contrary, the pairwise linguistic evaluations give information only on each single pair of actions. Thus both together can help decision-maker(s) in reaching a final decision.

We may conclude that since in a fuzzy environment a high precision of the results is illusory, the above procedure - aimed at supplying the decision-maker(s) with as much information as possible and at making the entropy levels connected with these evaluations as small as possible - is a meaningful undertaking.

7.7   Conclusions

The model developed in the present Chapter, the so-called NAIADE method, is a discrete multicriteria method whose impact (or evaluation) matrix may include either crisp, stochastic or fuzzy evaluations of the performance of an alternative a_n with respect to a criterion g_m. From an empirical point of view, this model is particularly suitable for economic-ecological modelling incorporating various degrees of precision of the variables taken into consideration. From a methodological point of view, two main issues are then faced:
- the problem of equivalence of the procedures used in order to standardise the various evaluations (of a mixed type) of the performance of alternatives according to different criteria;
- the problem of comparison of fuzzy numbers typical of all fuzzy multicriteria methods.

Since in a fuzzy context, any attempt to reach a high degree of precision on the results tends to be somewhat artificial, a pairwise linguistic evaluation of alternatives is used. This is done by means of the notion of fuzzy relations and linguistic quantifiers. In the aggregation process, particular attention is paid to the problem of diversity of the single evaluations, while the entropy concept is used as a measure of the associated "fuzziness". Such linguistic evaluations can be used in different ways according to the decision environment at hand.

In short, the main properties of the NAIADE method can be synthesised as follows:
- communication with the decision-maker is required to elicit different relevant parameters, thus a constructive decision aid framework is implied;
- the method is based on some aspects of the partial comparability axiom, in particular, a pairwise comparison between alternatives is carried out, and incomparability relations are allowed;

- intensity of preference is taken into account, this implies that a certain degree of compensation between criteria is allowed, given the characteristics of the method (in particular the parameter α in (7.17)) it may be classified among partial compensatory methods;
- for the indifference relation no transitivity is implied, the preference relation is max-min transitive;
- a partial (or total) order of feasible alternatives is supplied, thus the γ problem formulation is tackled. It has to be noted that the final ranking is a function of all the alternatives considered, this implies that if a dominated or a dominating action is introduced, the ranking may change; moreover if the best action is eliminated, the ranking of the other alternatives may also change, thus NAIADE does not respect the independence of irrelevant alternatives axiom.

In the NAIADE approach, a weighting of criteria is not assumed and no consideration is given to the "minority principle" (like the discordance index in the ELECTRE methods). Since in environmental and resource management and policy aiming at an ecologically sustainable development many conflicting issues and interests emerge, particular attention has to be given to the problem of different values and goals of different groups in society. Equity and conflicting values in multicriteria decision aid are traditionally introduced in two different ways:

(1)   by weighting the different criteria, but often in public decision making a single point-value solution (e.g. weights) tends to lead to deadlocks in a decision process because it imposes too rigid conditions for a compromise;

(2)   by taking into consideration a set of ethical evaluation criteria. A weak point of this approach is that it could lead to an excessive number of evaluation criteria, furthermore, to identify ethical criteria may be not an easy task.

The next Chapter, will deal with a third possibility i.e. the use of conflict analysis procedures to be integrated with multicriteria evaluation in order to allow policy-makers to seek for "defendable" decisions that could reduce the degree of conflict (in order to reach a certain degree of consensus) or that could have a higher degree of equity on different income groups.

APPENDIX 1
Membership Functions of Some Fuzzy Relations

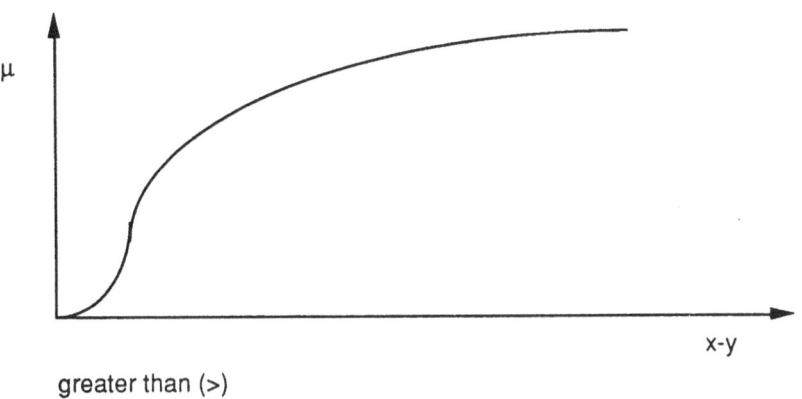

greater than (>)

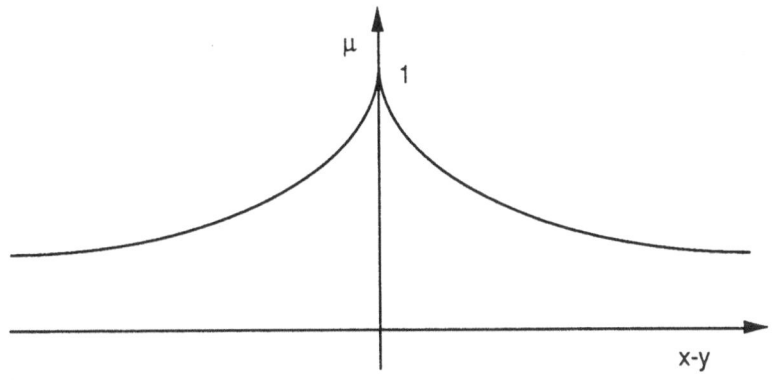

approximately equal to (≡)

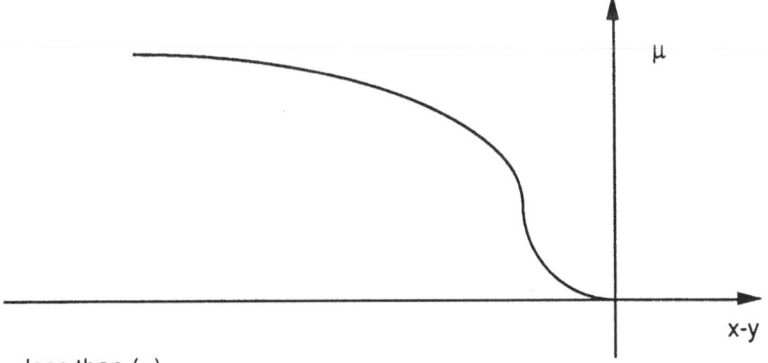

less than (<)

APPENDIX 2
Membership Functions of Some Linguistic Variables

In this Appendix the membership functions of the linguistic variables excellent, good, moderate and bad used in the illustrative example will be specified. We use the definitions by Li and Liu [1990].

$$\mu_{excellent}(x) = 1 - e^{-\left(\frac{0.5}{|1-x|}\right)^{2.5}}$$

$$\mu_{good}(x) = 1 - e^{-\left(\frac{0.25}{|0.4-x|}\right)^{2.5}}$$

$$\mu_{moderate}(x) = 1 - e^{-5|x|}$$

$$\mu_{bad}(x) = 1 - e^{-\left(\frac{0.25}{|-0.4-x|}\right)^{2.5}}$$

CHAPTER 8

A FUZZY CLUSTER PROCEDURE FOR ENVIRONMENTAL CONFLICT ANALYSIS

8.1   Conflicts in Environmental Management Problems

Generally, ecosystems are used in several ways at the same time by a number of different users. Such situations lead almost always to conflicts of interest and damage to the environment. Thus, in the area of environmental and resource management and in policies aiming at an ecologically sustainable development, many conflicting issues and interests emerge.

In Chapter 3, it has been shown that the *planning balance sheet method* [Lichfield, 1964, 1988, 1993] aims at providing a broader framework for the assessment of gains and losses of a plan by constructing detailed socio-economic accounts of all project effects and by taking into account different groups in society which are affected in their well-being by the plan. In this way both the efficiency and equity aspects can be considered. A weak point of this method is that it is primarily meant to present in a systematic way a description of all the distributive impacts, but no elaboration with normative purposes is generally made. A possible way to overcome this drawback of the planning balance sheet method is to integrate its philosophy with some results of coalition formation theory.

We will first review traditional literature on coalition formation theory, then a fuzzy clustering procedure based on both the philosophies of the planning balance sheet method and coalition formation theory will be presented.

8.2   Coalition Formation Theory: a Concise Overview

One of the aims of coalition formation theory is to predict a set of coalitions which are likely to be formed in a given political situation. There are two basic schools of thought among those who have applied game-theoretic principles to the study of political coalition formation. The two opposing positions can be referred to as "size theory" and "policy theory" [Holler, 1984].

Size theory originated from Von Neumann/Morgenstern's [1947] "minimal winning coalitions", and was modified by Riker [1962] and Gamson [1962] into "the size principle". Size theorists assert that parties prefer governments of which they are a member and which are "as small as possible". Size theorists argue that when a government coalition is voted into office it thereby gains control over a fixed sum of benefits which are then subdivided among its constituent members (with non-members receiving nothing). Therefore, the smaller the coalition, the more benefits are available per member.

"Policy theorists" such as Leiserson [1966], and De Swaan [1973] argue that the benefits to the political parties which are generated by a particular government come primarily from the policies implemented by the government. Since government policies are public goods, the benefits that a party may receive from different governments are not necessarily related to the size of the governments. Instead, these benefits are related to the preferences of that party for the policies of one government compared with those of the other.

There are two major variants of "policy theory": *"Minimal range theory"* and *"Policy distance theory"* [Holler, 1984].

Minimal range theory asserts that a particular party will prefer a government coalition of which it is a member and which has a small "range". Range can be defined as the distance in the policy space between the policy positions of the two most extreme members of a coalition. The argument underlying the minimal range hypothesis is that the smaller the range of the government, the closer the government policy is likely to be to the policy position of any one of its members. The minimal range hypothesis predicts government coalitions whose range is as small as possible, given that they must form a majority.

De Swaan [1973] makes the assumption that a coalition government selects its policies by a majority rule. For instance, thinking of the policy space as a line, this assumption leads to the conclusion that the policies chosen by a particular coalition government will be the policies of the party that is at the median of the coalition. According to the policy distance hypothesis, a party will prefer to belong to a coalition for which the policy position of the median member is close to its own position.

In *policy distance theories*, the construction of the predicted set of coalitions proceeds in two distinct steps: first, each actor establishes his preference ordering among the various possible coalitions, and then individual preferences are used to select a subset of all possible coalitions, i.e. the set of predicted coalitions. In general, the results of policy distance theory are less clear cut than minimal range theory, because it tends to predict that any of a relatively large number of coalitions is possible in a given period, whereas minimal range theory predicts a more restricted set of possibilities.

Some formal definitions will be given below.

Let N={1, 2,...,n} be a set of *n actors*. Let w_i, i=1, 2,...,n (w_i>0 $\forall$ i) be the weight of actor i, e.g. his number of seats in a parliament.

A *coalition S* is any subset of N. All coalitions are possible, including the empty coalition $\varnothing$, the grand coalition N and the singleton coalition {i}, i=1, 2, ... ,n. Let C denote the set of all possible coalitions and let C_i={S$\in$C:i$\in$S} i=1, 2,...,n be the set of coalitions of which actor i is a member. Then the *weight w_S of a coalition S* is the sum of the actors' weights which are members of S:

$$w_S = \sum_{i \in S} w_i \qquad \forall \, S \in C \tag{8.1}$$

A *majority criterion m* is the minimum number of votes sufficient to adopt a proposition.

In order to avoid two disjoint coalitions S and T (S$\cap$T=$\varnothing$) each with a majority, it is necessary to impose:

$$m > 1/2 w_N \tag{8.2}$$

A *winning coalition* is a coalition whose weight is greater than or equal to the majority criterion m, i.e.:

$$S \in C \text{ is winning if and only if } w_S \geq m \tag{8.3}$$

Let W={S$\in$C:$w_S$$\geq$m} be the *set of winning coalitions* and W_i denote the set of winning coalitions of which actor i is a member:

$$W_i = C_i \cap W = \{S \in W : i \in S\} = \{S \in C_i : S \in W\} \tag{8.4}$$

A voter i is an *essential member* of coalition W, if his defection from voting prevents the coalition value of W.

From a game-theoretical point of view, the following two behavioural assumptions are usually made:

Preference for membership of winning coalitions: an actor's objective is to become a member of a winning coalition, i.e.,

$$\forall \, i \in N, \, \forall \, S \in W_i, \, \forall \, T \notin W_i, \text{ it is } S P_i T \tag{8.5}$$

(where P_i indicates the preference relation according to actor i).

Preference for closer coalitions: among winning coalitions of which an actor is a member, he prefers coalitions whose expected policy position is perceived by him as being close to his own most preferred policy.

In order to compute such a distance, the following assumption can be made: any winning coalition in which an actor has more power, is placed by this actor closer to him than any other coalition.

If we are interested in determining the *a priori power* of a player in an N-person game, we must calculate the probability of the player being in a winning coalition. In turn, this requires us to determine the probability of each coalition taking place.

8.3   Some Principles of Game Theory

The cornerstone of the theory of *co-operative n-person games* [Moulin, 1988, Shubik, 1983] is the *characteristic function*. The idea is to capture in a single numerical index the potential worth of each coalition of players (the representation of coalition's worth by a single number implies freely transferable utility). Games are defined *inessential* if no profitable grounds exist for co-operation among players. They are *essential* if some members of coalitions at least, do strictly better by sticking together. Mathematically, the characteristic function, traditionally denoted by v, is a function from subsets of players to the set of real numbers. A general property of characteristic functions is that the function v is *super-additive* (any set of players can do at least as well in coalition as in any sub-coalition), formally,

$$v(S \cup T) \geq v(S) + v(T) \tag{8.6}$$

which holds whenever S and T have no members in common.

A game is defined as a "*c-game*" if it can adequately be represented by its characteristic function (nothing essential to the ultimate purpose of the model is lost in the process of condensing the extensive or strategic description into a characteristic function). In general, two classes of, games qualify as c-games, no matter what solution concept is adopted. These games are:

•   constant-sum games, in which the total pay-off is a fixed quantity, regardless of the strategy chosen;

•   games of consent or games of orthogonal coalitions, the idea here, is that nothing can happen to change a player's pay-off unless he himself is a party in the action (either you cooperate with someone or you ignore him; you cannot actively hurt him). In economic theory, this is the case of models of pure competition without externalities.

A measure game is a game in which the worth of a coalition is determined by the measure of some resource that the coalition members can pool.

A special class of games, which may be thought of as "games of control" are important tools in the modelling of organisational and group decision processes. They are called *simple games* and are distinguished by the property of having just two kinds of coalitions, namely, winning and losing. In the presence of transferable utility, they are c-games, and after suitable normalisation they give rise to a special single type of characteristic function:

$$v(S)= \begin{cases} 0 & \text{if S is losing} \\ 1 & \text{if S is winning} \end{cases} \qquad (8.7)$$

The theory of simple games is often presented without reference to a characteristic function. The usual axioms are:
- every coalition is either winning or losing;
- the empty coalition is losing;
- the all-player set is winning;
- no losing set contains a winning subset.
  An additional axiom is required for superadditivity:
- the complement of any winning set is losing; a game satisfying this axiom is said to be *proper*. Such an axiom prevents the confusion that might result from allowing separate winning coalitions to make simultaneous decisions. Yet another axiom is needed to make the game constant sum:
- the complement of any losing set is winning; a game satisfying this axiom is said to be *strong*. Such an axiom prevents the paralysis that might result from allowing a losing coalition to obstruct the decision.

In general, strong proper games, also called *decisive games*, represent efficient group decision rules, whereas the other simple games represent procedures that may be plagued by inconsistencies, deadlocks, or both.

8.4 Power Indices in Comparison

Two main power indices can be found in the literature:

- the Shapley/Shubik power index, and
- the Banzhaf/Coleman power index.

8.4.1 The Shapley/Shubik Power Index

This power index is an alternative interpretation of the so-called Shapley value. The Shapley value falls into the case of c-games with transferable utility; when an n-person game is presumed to be adequately represented by an ordinary numerical characteristic function, then the value pay-offs to the several players can depend on only 2^n-1 numbers $v(S)$, as S ranges over all non-empty subsets of the all-player set. Is there any way we can deduce the functional form of this dependence? Given some axioms, Shapley showed that an exact value exists, i.e.

$$\Phi_i = \frac{1}{n} \sum_{s=1}^{n} \frac{1}{c(s)} \sum_{\substack{S \ni i \\ |S|=s}} [v(S) - v(S - \{i\})] \tag{8.8}$$

where $c(s)$ is the number of coalitions of size s containing the designated player i:

$$c(s) = \binom{n-1}{s-1} = \frac{(n-1)!}{(n-s)!(s-1)!} \tag{8.9}$$

This formula means that the value of the game to a player is his average marginal worth to all coalitions in which he might participate. This value solution has an interesting interpretation when applied to simple games. Most of the marginal worthies $[v(S)-v(S-\{i\})]$ are zero; only when a coalition is very close to the borderline between winning and losing will the adherence or defection of a single member make any difference. The value then, is in effect a way of counting the opportunities a player has of being pivotal in the group decision process. In any ordering of the players the coalition that is being built changes at some point from losing to winning. The player who effects this change is called the *pivot* of that ordering. The Shapley value to any player is precisely the probability that he will be the pivot. Therefore, if this solution is interpreted as a probability rather than as a share of a divisible prize, then it too is virtually freed from utility considerations and becomes not a value of the game but a dimensionless measure of the distribution of power among the players. In this interpretation the value solution is called the Shapley/Shubik [1954] power index.

The empirical relevance of this index has been questioned because of the underlying concept of *equiprobable permutations* of N. This concept has been justified on the grounds that *since neither the issues nor the preferences of the actors are specified*, we can ascribe to the principle of insufficient reason equal probability to any issue and corresponding preference ordering, i.e. permutations of voters.

8.4.2 The Banzhaf/Coleman Power Index

The key element in the construction of the Banzhaf [1965] index is a *"swing"*. A swing for player i is a pair of sets (S, S-{i}) such that S is a winning set and S-{i} is not. In other words, a swing occurs when the defection of voter i changes a coalition from winning to losing. For each player i in the set of all players N, we may calculate a number $\eta_i(v)$ that represents the numbers of swings for i in the simple game with characteristic function v. Let $\eta'(v)$ be the total number of swings in the game,

$$\eta'(v) = \sum_{i \in N} \eta_i(v) \tag{8.10}$$

a player with $\eta_i(v)=0$ is called a dummy since he is never needed to help a coalition win. A player with $\eta_i(v)=\eta'(v)$ is called a dictator. In order to have an index of relative voter power, it is useful to define a normalised Banzhaf index such that the weights add to one:

$$\beta_i(v) = \frac{\eta_i(v)}{\eta'(v)} \qquad (i=1, 2,....,n) \tag{8.11}$$

Some problems connected with this normalised form were pointed out and an alternative normalisation has been suggested [Shubik, 1983]:

$$\beta'_i(v) = \frac{\eta_i(v)}{2^{n-1}} \qquad (i=1, 2,....,n) \tag{8.12}$$

that can be interpreted as the *probability that a player is a swinger*.
Coleman [1977] have proposed a power index that is a linear transformation of the Banzhaf index. Thus often it is called the Banzhaf/Coleman Power Index.

8.4.3 Comments

In general, the numerical results of the Shapley/Shubik index and the Banzhaf index agree quite well. They diverge seriously for "near-oceanic" games (games with several powerful players and a large number of weak players). If the game can be presented as a voting body, the Shapley/Shubik index and the Banzhaf/Coleman index are monotonic with respect to votes; both indices are however not necessarily monotonic with respect to changes in voting weights.

Shapley/Shubik index depends on *equiprobable permutations of N*, whereas the Banzhaf index depends on *equiprobable combinations* of N. Each permutation produces exactly one pivot, but a single combination rarely produces a single swinger. Thus there is an inherent additive measure in the Shapley/Shubik index which is not present in the Banzhaf index. An interesting comparison of the characteristics of power indices can be found in Gambarelli [1983].

8.5   Fuzzy Cluster Analysis in Coalition Formation Theory

Given these considerations, it is possible to construct a model (performing as a "simulation model") whose main aim is to give relevant information on the structure of the decision problem at hand. For example, the authority in charge of a decision can try to forecast the possible behaviour of the relevant interest groups so that actions, having a higher probability of being accepted by certain groups, may be identified; or in the framework of an equity analysis, it is possible to evaluate the impact of different actions on different income/interest groups.

As in the planning balance sheet method, this approach requires the construction of a matrix showing the various alternatives and the impacts of these alternatives on different income groups.

The following main assumptions are made:

(1) only a set of well defined actions has to be taken into account;

(2) the impacts of these alternatives on different income/interest groups are evaluated by means of "linguistic declarations" (good, not very good, etc.);

Given a conflict indicator, a fuzzy cluster algorithm can be used in order to have an idea of the coalitions (minimising such an indicator) that are "possible". It should be noted that the formal structure of the model is:

units=interest/income groups

attributes=actions.

Thus we have to evaluate the similarity among groups given the evaluations of the different actions. By using the distance described in (6.70) as conflict indicator, a similarity matrix (achieved by means of the simple transformation s=1/1+d) for all possible pairs of groups can be obtained, so that the following clustering procedure is meaningful.

On an axiomatic basis, cluster analysis can be distinguished in deterministic, stochastic and fuzzy. By taking into consideration the "clustering criteria", the following distinction exists [Anderberg, 1973, Bezdek, 1980, Hartigan, 1975]:

- hierarchical methods,
- graph theoretic methods,
- objective functional methods.

The hierarchical clustering approach, in particular, allows an evolutionary view of the aggregation process and can easily be dealt within fuzzy terms.

However, in a fuzzy environment a problem exists, i.e. the relation between the concepts of *partition* and *equivalence class*. In a crisp environment, the choice of treatment of data in terms of partitions or equivalence relations is a matter of convenience, since the two models are fully equivalent (philosophically and mathematically). On the contrary, fuzzy equivalence relations and partitions are philosophically similar, but their mathematical structures are not isomorphic (e.g. the notion of transitivity is unique for crisp relations but has taken several proposed forms in the fuzzy case).

We start the discussion on fuzzy cluster analysis with the definition of a *crisp equivalence relation*.

Let $X=\{x_1, x_2,....,x_n\}$ be any finite set. Then an nxm matrix $R=[r_{ij}] = [r(x_i, x_j)]$ is a crisp equivalence relation on XxX if

$r_{ii}=1$	$1 \leq i \leq n$	(reflexivity)	(8.13)
$r_{ij}=r_{ji}$	$1 \leq i \neq j \leq n$	(symmetry)	(8.14)

$$\left\{ \begin{array}{l} r_{ij}=1 \\ r_{jk}=1 \end{array} \right. \Rightarrow r_{ik}=1 \quad \forall i,j,k \qquad \text{(transitivity)} \qquad (8.15)$$

Let R be a fuzzy binary relation with $\mu_R(x_i, x_j)$ indicating the degree to which two elements x_i and x_j are similar (similarity matrix). The relation R is obviously reflexive and symmetric, thus it is called a *resemblance relation*.

A fuzzy relation is a *similitude relation* if it has the following properties:

$$\mu_R(x_i, x_i) = 1 \qquad \forall \, (x_i, x_i) \in X \times X \qquad \text{(reflexivity)} \qquad (8.16)$$

$$\mu_R(x_i, x_j) = \mu_R(x_j, x_i) \qquad \forall \, (x_i, x_j) \in X \times X \qquad \text{(symmetry)} \qquad (8.17)$$

$$\mu_R(x_i, x_k) \geq \max \min [\mu_R(x_i, x_j), \mu_R(x_j, x_k)]$$

$$\forall \, (x_i, x_j), (x_j, x_k), (x_i, x_k) \in X \times X \qquad \text{(max-min transitivity)} \qquad (8.18)$$

Note that compared to the notion of transitivity in conventional analysis, the present notion defines a weak transitivity of similarity.

If one wants to derive a set of equivalence classes (and not simple partitions) there is a need for the similarity matrix being at least max-min transitive. As it is known [Leung, 1988], a method to transform an intransitive similarity matrix into a transitive one is to derive the *transitive closure* $\mathcal{R}$ of R. The *max-min transitive closure* of a fuzzy binary relation R is

$$\mathcal{R} = R \cup R^2 \cup R^3 \cup \ldots \qquad (8.19)$$

where $R^2 = R \circ R$ is the max-min composition of R.

The *max-min composition* is a standard operation for two fuzzy relations: given two relations R(x, y), S(y, z) defined on $X \times Y$ and $Y \times Z$, respectively, the max-min composition of R and S, denoted as $R \circ S$, is defined by

$$\mu_{R \circ S}(x, z) = \max_{y \in Y} \min [\mu_R(x, y), \mu_S(y, z)] \qquad (8.20)$$

$x \in X$, $y \in Y$ and $z \in Z$.

By using the notion of max-min composition, one is allowed to derive new fuzzy relations. A transitive closure can be obtained by means of the following theorem [Leung, 1988].

Let R be any fuzzy binary relation. If for some k, the max-min composition $R^{k+1} = R^k$, then the max-min transitive closure is

$$\mathcal{R} = R \cup R^2 \cup R^3 \cup \ldots \cup R^k \qquad (8.21)$$

Knowing that a fuzzy set A can always be decomposed into a series of α-level sets A_α, the similitude relation $\mathcal{R}$ can be decomposed into

$$\mathcal{R} = \bigcup_{\alpha \in [0,1]} \alpha \cdot \mathcal{R}_\alpha \qquad (8.22)$$

where

$$f_{\mathcal{R}}(x_i, x_j) = \begin{cases} 1 & \text{if } \mu_{\mathcal{R}}(x_i, x_j) \geq \alpha \\ 0 & \text{if } \mu_{\mathcal{R}}(x_i, x_j) < \alpha \end{cases} \qquad (8.23)$$

and $\alpha_1 > \alpha_2 \Rightarrow \mathcal{R}_{\alpha_1} \subset \mathcal{R}_{\alpha_2}$ $\qquad (8.24)$

Since $\mathcal{R}_\alpha$ is reflexive, symmetric and transitive in the sense of ordinary sets, then it is an *equivalence class of level* α. Within each α-level equivalence class, the similarity of any two units is no less than α.

Note that the equivalence classes obtained are ordinary disjoint sets. In fact, in order to have non-mutually exclusive equivalence classes, it is necessary to assume the use of a *min-addition transitive distance matrix* (which is a stronger assumption than max-min transitivity).

It can be proved that the following four algorithms generate the same partition [Miyamoto, 1990]:
• the single linkage method,
• the connected components of an undirected fuzzy graph,
• the transitive closure of a reflexive and symmetric fuzzy relation, and
• the maximal spanning tree of a weighted graph.

Then the following conclusions can be drawn:
• since the connected components are independent of the numbering of the vertices, the algorithm is independent of the ordering of the inputs, and therefore it is *stable*;
• no *reversal* exists in the dendrogram (reversal meaning that the merging levels are not monotonically decreasing, and thus a cut of the dendrogram may produce ambiguous results);
• one is not obliged to use only the Euclidean metric (e.g. like in the "centre of gravity" procedures), but any distance measure (even if it does not respect the triangular inequality property) can be used; thus the method is quite *general*.

This Chapter has been devoted to the role of conflict analysis in environmental management. A fuzzy conflict resolution procedure aimed to be integrated with the NAIADE method has been presented. As an illustrative example, the whole procedure will be applied to an environmental management problem faced in The Netherlands some years ago.

8.6 Multicriteria Evaluation and Fuzzy Conflict Analysis: Illustration by Means of a Land Use Problem

The illustrative application used in this section is based on a earlier case study which was using ordinal information and multidimensional scaling techniques [Nijkamp, 1980]. It concerns a study on environmental management in the Netherlands (the southern part of Limburg). The problem at hand can be briefly illustrated as follows.

A company having almost absolute dominance in the Dutch cement industry has a concession to extract marl on one of the hills in south Limburg, but this concession may finish in the near future; thus alternative areas have to be explored. Among the new possible areas, the most appropriate one is the Plateau van Margraten; this is a rather flat area which is used for agriculture and for some recreation. It has a unique physical structure and it is a rather characteristic area in the landscape of the region. Designation of this area for marl winning would fundamentally affect its social and ecological value; on the other hand, if the authorities would refuse to grant permission for marl winning to the company, this would lead to an almost total destruction of the national cement industry and to serious unemployment effects for this already weak economic region. This situation clearly demonstrates the sharp conflicts between environmental and economic interests.

A first meaningful step toward an evaluation analysis for this land use problem is to identify a set of feasible and relevant alternatives. These alternatives are:

(1) An implementation of the original plans of the company (i.e. a concession for the total area). This guarantees the future position of the national cement industry and also favours the employment and welfare in the region. Agriculture suffers from some negative impacts, while the negative social impacts (for recreation, etc.) are rather high. Finally, the environmental damage is very high.

(2) The use of an alternative area (the Rasberg area, in the same region) for marl winning. But this area is much smaller and the physical condition of

the soil hampers a profitable cement production against current prices. On the other hand, the ecological damage is less serious.

(3) The provision of a concession for one half of the area (Plateau van Margraten). This leads to less agricultural losses, while the environmental damage is also lower. The economic impacts are less favourable than those of the first alternative.

(4) A new concession for marl winning on the present area. This is only a short-term solution which is less attractive from an economic point of view (note that in this case, one would need a multi-period approach, but this is too complex for illustrative purposes).

(5) Import of marl from the Plateau van Vroenhoven, an area in Belgium. This solution may be attractive from a social and environmental point of view (at least from a national stand-point), but it is less attractive from an economic point of view. For simplicity, we ignore the environmental impact of transport of marl.

(6) A restructuring of the company so that it becomes a trade and research organisation for cement instead of a production unit for cement. This will lead to a certain loss of employment, while the future need for such an organisation is unclear.

(7) A close-down of all productive activities of the company. This may be favourable from the viewpoint of environmentalists and recreationers, but it will lead to serious economic problems for the region.

These alternatives are to be judged on the basis of various evaluation criteria. Three main groups of criteria can be distinguished, viz. economic, social and environmental. These three classes can be subdivided into various components.

A)   Economic Criteria
1)   employment in agriculture,
2)   employment in cement industry (including marl winning),
3)   agricultural production,
4)   national production of marl,
5)   value added in cement industry

B)   Social Criteria
6)   residential attractiveness,
7)   recreational attractiveness (daily),
8)   tourist attractiveness,
9)   congestion created in transportation infrastructure.

C) Environmental Criteria

10) quality of physiological structure,
11) diversity and scarcity of eco- and bio-components,
12) consistency with existing landscape components,
13) consistency with existing cultural-historical components.

It appears that the information concerning the diverse plan impacts is rather inaccurate; the degree of uncertainty on the impacts of the plans is high, so that quantitative information on these impacts is often not available. A representation of such impacts in fuzzy terms seems very appropriate. A multicriteria fuzzy evaluation matrix related to the above-mentioned 7 alternatives and 13 criteria is presented in Table 8.1.

In addition to this fuzzy evaluation matrix, an assessment of the priority structures of the diverse interest groups is required. The number of interest groups distinguished in this study was six. These groups are:

1) the board of directors of the company,
2) the employees of the company,
3) the farmers' association in Limburg,
4) the recreational association for South Limburg,
5) the environmental federation in Limburg,
6) the residents of the area around the Plateau Margraten.

Criteria	Alternatives						
	a_1	a_2	a_3	a_4	a_5	a_6	a_7
g_1	moderate	moderate	good	good	excellent	excellent	excellent
g_2	excellent	excellent	moderate	moderate	good	moderate	bad
g_3	moderate	moderate	good	good	excellent	excellent	excellent
g_4	excellent	excellent	moderate	moderate	moderate	bad	bad
g_5	excellent	moderate	bad	bad	good	good	bad
g_6	moderate	moderate	bad	bad	good	good	bad
g_7	good	good	good	excellent	excellent	excellent	excellent
g_8	moderate	moderate	good	excellent	excellent	excellent	excellent
g_9	moderate	moderate	moderate	excellent	excellent	excellent	excellent
g_{10}	moderate	moderate	moderate	excellent	excellent	excellent	excellent
g_{11}	good	good	good	excellent	excellent	excellent	excellent
g_{12}	bad	moderate	good	excellent	excellent	excellent	excellent
g_{13}	moderate	good	good	excellent	excellent	excellent	excellent

Table 8.1 Evaluation Matrix for a Fuzzy Land Use Problem

Interest groups	Alternatives						
	a_1	a_2	a_3	a_4	a_5	a_6	a_7
1	very good	good	moderate	bad	fairly good	fairly bad	very bad
2	very good	good	moderate	bad	fairly good	very bad	very bad
3	very bad	fairly bad	moderate	good	very good	good	moderate
4	very bad	fairly bad	fairly bad	good	fairly good	good	very good
5	very bad	bad	fairly bad	moderate	fairly good	good	very good
6	very bad	good	bad	good	good	good	very good

Table 8.2 Fuzzy Evaluations of Alternatives According to Each Interest Group

In Table 8.2 the linguistic evaluations of the alternative plans according to each interest group are presented. These evaluations were assessed on the basis of personal inquiries, interviews, talks with interest groups and study of available material.

By applying NAIADE, for each pair of actions, the following degrees of truth (expressing the credibility of the pairwise comparison results) of a linguistic evaluation are obtained (for $\alpha=0.40$):

a_1 is better than a_2	$\tau=0$		a_1 is better than a_3	$\tau=0$
a_1 and a_2 are indifferent	$\tau=1$		a_1 and a_3 are indifferent	$\tau=0$
a_1 is worse than a_2	$\tau=0$		a_1 is worse than a_3	$\tau=0$

a_1 is better than a_4	$\tau=0$		a_1 is better than a_5	$\tau=0$
a_1 and a_4 are indifferent	$\tau=0$		a_1 and a_5 are indifferent	$\tau=0$
a_1 is worse than a_4	$\tau=1$		a_1 is worse than a_5	$\tau=1$

a_1 is better than a_6	$\tau=0$		a_1 is better than a_7	$\tau=0$
a_1 and a_6 are indifferent	$\tau=0$		a_1 and a_7 are indifferent	$\tau=0$
a_1 is worse than a_6	$\tau=1$		a_1 is worse than a_7	$\tau=1$

a_2 is better than a_3	$\tau=0$		a_2 is better than a_4	$\tau=0$
a_2 and a_3 are indifferent	$\tau=0$		a_2 and a_4 are indifferent	$\tau=0$
a_2 is worse than a_3	$\tau=0$		a_2 is worse than a_4	$\tau=1$

a_2 is better than a_5	$\tau=0$		a_2 is better than a_6	$\tau=0$
a_2 and a_5 are indifferent	$\tau=0$		a_2 and a_6 are indifferent	$\tau=0$
a_2 is worse than a_5	$\tau=1$		a_2 is worse than a_6	$\tau=1$

a_2 is better than a_7	$\tau=0$		a_3 is better than a_4	$\tau=0$
a_2 and a_7 are indifferent	$\tau=0$		a_3 and a_4 are indifferent	$\tau=0$
a_2 is worse than a_7	$\tau=1$		a_3 is worse than a_4	$\tau=1$

a_3 is better than a_5	$\tau=0$		a_3 is better than a_6	$\tau=0$
a_3 and a_5 are indifferent	$\tau=0$		a_3 and a_6 are indifferent	$\tau=0$
a_3 is worse than a_5	$\tau=1$		a_3 is worse than a_6	$\tau=1$

a_3 is better than a_7	$\tau=0$		a_4 is better than a_5	$\tau=0$
a_3 and a_7 are indifferent	$\tau=0$		a_4 and a_5 are indifferent	$\tau=1$
a_3 is worse than a_7	$\tau=1$		a_4 is worse than a_5	$\tau=0$

a_4 is better than a_6	$\tau=0$		a_4 is better than a_7	$\tau=0$
a_4 and a_6 are indifferent	$\tau=1$		a_4 and a_7 are indifferent	$\tau=1$
a_4 is worse than a_6	$\tau=0$		a_4 is worse than a_7	$\tau=0$

a_5 is better than a_6	$\tau=0$		a_5 is better than a_7	$\tau=0$
a_5 and a_6 are indifferent	$\tau=1$		a_5 and a_7 are indifferent	$\tau=1$
a_5 is worse than a_6	$\tau=0$		a_5 is worse than a_7	$\tau=0$

a_6 is better than a_7	$\tau=0$
a_6 and a_7 are indifferent	$\tau=1$
a_6 is worse than a_7	$\tau=0$

It is clear that almost all linguistic evaluations are quite unambiguous. This is caused by four factors:
(1) the number of criteria in favour of an action;
(2) the degree of compensation allowed in the aggregation process ($\alpha=0.40$);
(3) definition of the membership function of the linguistic operators;
(4) aggregation operator chosen for the approximate reasoning operations (in this application we have used the "min" operator).

It has to be noted that between actions a_1 and a_3, and a_2 and a_3, none of the possible situations satisfies the minimum requirement requested by the linguistic operator "most"; this can be interpreted as a difficulty in the comparison which might lead to an incomparability relation.

On the basis of the above pairwise comparison between alternatives, we arrive at the following final ranking of alternatives:

$$\{a_4, a_5, a_6, a_7\} > \{a_1, a_2, a_3\}.$$

This means that we obtain two subsets of alternatives. The best subset contains a_4, a_5, a_6 and a_7. On the basis of the pairwise comparison results, it is not possible, however, to rank the alternatives within the two subsets.

An higher degree of discrimination can be obtained by computing the ϕ^+ and ϕ^- indices; the following results are obtained:

$$a_6 \rightarrow a_5 \rightarrow a_4 \rightarrow a_7 \rightarrow a_1 \rightarrow a_2$$
$$\searrow$$
$$a_3$$

This (ordinal) ranking appears quite stable. In fact, only two incomparability relations are present, i.e. between a_1 and a_3, and between a_2 and a_3. Finally, it has to be noted that this ranking includes a higher degree of discrimination than the pairwise linguistic evaluations (e.g. actions a_1 and a_2 are not anymore considered indifferent), because it is a function of the relation of each alternative with all other ones, and no "minimum requirement threshold" like in the linguistic quantifiers is introduced.

However, since a weighting of criteria is not assumed and no consideration is given to the "minority principle" (like the discordance index in the ELECTRE methods) such a multicriteria procedure must be integrated with conflict minimisation methods which allow policy-makers to seek for "defendable" decisions that could reduce the degree of conflict (in order to reach a certain degree of consensus) or that could have a higher probability of being accepted by certain interest groups.

Taking into consideration the possibility of coalitions among the different interest groups, the following results can now be obtained.

By applying the semantic distance described in (6.70) with p=2, after the transformation s= 1/1+d, the following similarity matrix for all possible pairs of interest groups is obtained:

	1	2	3	4	5	6
1	1	0.729	0.426	0.399	0.403	0.403
2	0.729	1	0.410	0.386	0.390	0.390
3	0.426	0.410	1	0.675	0.584	0.569
4	0.399	0.386	0.675	1	0.729	0.672
5	0.403	0.390	0.584	0.729	1	0.595
6	0.403	0.390	0.569	0.672	0.595	1

This means for example that the highest similarity occurs for interest groups 1 and 2, and for groups 4 and 5. These interest groups have a relatively high correspondence of goals, accordingly. The reverse holds true for interest groups 2 and 4 where the lowest degree of similarity is found.

By using the notion of max-min composition, the following new fuzzy relations are derived:

R^2

	1	2	3	4	5	6
1	1	0.729	0.426	0.426	0.426	0.426
2	0.729	1	0.426	0.410	0.410	0.410
3	0.426	0.426	1	0.675	0.675	0.672
4	0.426	0.410	0.675	1	0.729	0.672
5	0.426	0.410	0.675	0.729	1	0.672
6	0.426	0.410	0.672	0.672	0.672	1

R^3

	1	2	3	4	5	6
1	1	0.729	0.426	0.426	0.426	0.426
2	0.729	1	0.426	0.426	0.426	0.426
3	0.426	0.426	1	0.675	0.675	0.672
4	0.426	0.426	0.675	1	0.729	0.672
5	0.426	0.426	0.675	0.729	1	0.672
6	0.426	0.426	0.672	0.672	0.672	1

R^4

	1	2	3	4	5	6
1	1	0.729	0.426	0.426	0.426	0.426
2	0.729	1	0.426	0.426	0.426	0.426
3	0.426	0.426	1	0.675	0.675	0.672
4	0.426	0.426	0.675	1	0.729	0.672
5	0.426	0.426	0.675	0.729	1	0.672
6	0.426	0.426	0.672	0.672	0.672	1

Since in the series of max-min compositions $R^3 = R^4$, the transitive closure is
$$\mathcal{R} = R \cup R^2 \cup R^3 = R^3$$

Since $\mathcal{R}$ is a similitude relation, it can be decomposed into equivalence classes with respect to the degree of similarity α.

Thus the application of the clustering procedure leads to the following results.

As long as the similarity degree α required for a coalition is higher than .729, there will be no coalition formation. Two coalitions will be formed when α is between .729 and .675 (1 and 2), and (4 and 5). When the similarity degree is reduced to .675 and .672, interest groups 3 and 6 join the last coalition, respectively. The conflict of interest between the remaining coalitions (1, 2) versus (3, 4, 5, 6) is considerable as can be inferred from the low similarity degree associated with a grand coalition.

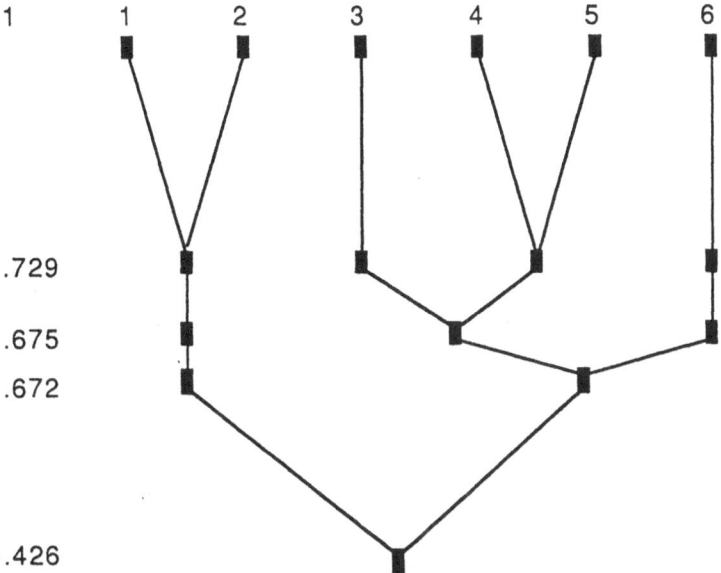

Figure 8.1   Dendrogram of the Coalition Formation Process

These results are mostly in agreement with prior expectations about the attitudes and behaviour of the interest groups. The interests of the company and of its employees seem to run fairly parallel. The agricultural interest group seems to take an intermediate position, but it joins quite soon the coalition made by the recreation and environmental groups. The priority patterns of the recreation group and the environmental group bear a very close correspondence. The residential group presents a more individualistic character since it can be considered a clear case of a "NIMBY" (never in my back yard) syndrome; in any case, it is closer to the interests of the recreation and environmental groups than to those of the economic groups.

It is interesting to note that the alternatives strongly supported by interest groups 1 and 2 (a_1, a_2) have bad environmental impacts. All the alternatives considered "good" from an environmental point of view are more or less well-accepted by interest groups 3, 4, 5, 6. Among the actions of this group, a_6, a_5, and a_4 are clearly compromise solutions in nature while a_7 is too extreme a solution (close down of all productive activities) but which clearly presents a high performance from a social and environmental point of view; a_5 is the only alternative which minimises the conflicts (this is clear by looking to Table 8.2). Both a_6 and a_4 will strongly be rejected by interest groups 1 and 2.

8.7 Concluding Remarks

In the area of environmental and resource management and in policies aiming at an ecologically sustainable development, many conflicting issues and interests are the normal state of affairs. Multicriteria evaluation techniques cannot solve all these conflicts, but they can help to provide more insight into the nature of these conflicts by providing systematic information and into ways to arrive at political compromises in case of divergent preferences in a multi-group or committee system by making the trade-offs in a complex situation more transparent to decision-makers.

In the present Chapter, the conflict issues have been taken into consideration by means of a fuzzy conflict analysis procedure aimed at being integrated with multicriteria evaluation. Starting with a matrix showing the impacts of different courses of action on each different interest/income group, a fuzzy clustering procedure indicating the groups whose interests are closer in comparison with the other ones is used. Therefore, finally a compromise solution taking into account all three conflictual values of economics (efficiency, equity and sustainability) can in principle be isolated.

Up till now a weak element in this conflict analysis is the lack of strategic considerations leading to new coalitions or alliances. In fact, the clustering algorithm only indicates the groups whose interests are closer in comparison to the other ones. This is more or less in agreement with the hypotheses underlying the "minimal range theory"; game-theoretic elements such as the notion of "power" need to be introduced. Furthermore, attaching to each interest group the same weight can be an oversimplification of a real-world situation.

Finally, one has to note that the results of NAIADE depend on different factors. Thus, Chapter 9 focuses on the analysis of the behaviour of such a fuzzy multicriteria method. This is achieved by means of sensitivity analysis.

CHAPTER 9

SENSITIVITY ANALYSIS IN THE NAIADE METHOD

9.1   Introduction

In a constructive decision aid framework, there is a need for an elicitation process requiring precise information. This process means that the decision-maker's judgements are encoded by parameters p, the space of parameters being P. However, it may be difficult to elicit the information to build them exactly. Hence, two not necessarily mutually exclusive, cases arise [Rìos Insua, 1990]:

- the decision-maker is not able to locate p precisely, but he may give some constraints on it (decision making under partial information);
- the decision-maker gives an estimate π of p, but, feeling doubtful about it, he asks for some help to identify which of his judgements are the critical ones in order to think about them more closely. In this case there is a need for sensitivity analysis.

In a multicriteria evaluation framework, different approaches to sensitivity analysis can be found in Rìos Insua [1990], French [1986], Belton and Vickers [1990], Janssen [1992], Thevenet and Rietveld [1992], Roy et al. [1986] (in particular, these last authors suggest a so-called robustness analysis approach).

In this Chapter, sensitivity analysis will be used to test the robustness of the results provided by the NAIADE method. In particular the role of the parameter α, used in the aggregation process, and of the aggregation operator used in the approximate reasoning operations will be investigated.

9.2   Sensitivity Analysis in the NAIADE Method: an Example

Let us take into consideration the simple example illustrated in Table 9.1:

	Alternatives			
Criteria	a	b	c	d
g_1	good	moderate	bad	excellent
g_2	bad	excellent	moderate	good
g_3	excellent	moderate	excellent	good
g_4	good	excellent	good	good

Table 9.1 A Fuzzy Evaluation Matrix

By applying the NAIADE method various pairwise "fuzzy preference relations" according to different levels of α are obtained (see equations (7.17) and (7.20)). As an example, here we will present only the ones referring to the pair **a**, **b**; the full set of outcomes can be found in Appendix 1.

$$
\left(
\begin{array}{ll}
\mu_{>>}(a, b)=1 & C(>)=0.587 \\
\mu_{>}(a, b)=1 & C(>>)=0.604 \\
\mu_{\cong}(a, b)=1 & C(\cong)=0.575 \\
\mu_{=}(a, b)=1 & C(=)=1 \\
\mu_{<}(a, b)=1 & C(<)=0.758 \\
\mu_{<<}(a, b)=1 & C(<<)=0.753
\end{array}
\right) \qquad (\alpha=0)
$$

$$
\left(
\begin{array}{ll}
\mu_{>>}(a, b)=0.642 & C(>)=0.587 \\
\mu_{>}(a, b)=0.705 & C(>>)=0.604 \\
\mu_{\cong}(a, b)=0.298 & C(\cong)=0.575 \\
\mu_{=}(a, b)=0 & C(=)=1 \\
\mu_{<}(a, b)=0.75 & C(<)=0.758 \\
\mu_{<<}(a, b)=0.736 & C(<<)=0.753
\end{array}
\right) \qquad (\alpha=0.20)
$$

$$
\left(
\begin{array}{ll}
\mu_{>>}(a, b)=0.309 & C(>)=0.815 \\
\mu_{>}(a, b)=0.411 & C(>>)=0.604 \\
\mu_{\cong}(a, b)=0 & C(\cong)=1 \\
\mu_{=}(a, b)=0 & C(=)=1 \\
\mu_{<}(a, b)=0.5 & C(<)=0.758 \\
\mu_{<<}(a, b)=0.47 & C(<<)=0.753
\end{array}
\right) \qquad (\alpha=0.40)
$$

$$
\left(
\begin{array}{ll}
\mu_{>>}(a, b)=0.114 & C(>)=0.815 \\
\mu_{>}(a, b)=0.116 & C(>>)=0.854 \\
\mu_{\cong}(a, b)=0 & C(\cong)=1 \\
\mu_{=}(a, b)=0 & C(=)=1 \\
\mu_{<}(a, b)=0.25 & C(<)=1 \\
\mu_{<<}(a, b)=0.27 & C(<<)=1
\end{array}
\right) \qquad (\alpha=0.60)
$$

$$
\left\{
\begin{array}{ll}
\mu_{>>}(a, b)=0 & C(>)=1 \qquad\qquad (\alpha=0.60) \\
\mu_{>}(a, b)=0.30 & C(>>)=0.854 \\
\mu_{\cong}(a, b)=0 & C(\cong)=1 \\
\mu_{=}(a, b)=0 & C(=)=1 \\
\mu_{<}(a, b)=0.1 & C(<)=1 \\
\mu_{<<}(a, b)=0.09 & C(<<)=1
\end{array}
\right.
$$

One can note that if α is equal to zero, no discrimination between actions is possible. When α increases, the value of the variable C also increases, on the contrary, the value of each fuzzy relation decreases. This behaviour is mainly due to the fact that when α increases, only values having a high intensity of preference or indifference are used. Moreover, when α increases, a lower degree of compensation among the criteria is allowed. In conclusion, one has to note that if values of α too low or too high are used, it is very difficult to discriminate between actions.

For each pair of actions, the following degrees of truth of a linguistic evaluation are obtained (see equations from (7.21) to (7.24)); here, the min operator is used.

a is better than b	$\tau=1$	
a and b are indifferent	$\tau=0$	$(\alpha=0.20)$
a is worse than b	$\tau=1$	

a is better than b	$\tau=0.028$	
a and b are indifferent	$\tau=0$	$(\alpha=0.40)$
a is worse than b	$\tau=0.474$	

a is better than b	$\tau=0$	
a and b are indifferent	$\tau=0$	$(\alpha=0.60)$
a is worse than b	$\tau=0$	

a is better than b	$\tau=0$	
a and b are indifferent	$\tau=0$	$(\alpha=0.80)$
a is worse than b	$\tau=0$	

a is better than c	$\tau=0.367$	
a and c are indifferent	$\tau=1$	$(\alpha=0.20)$
a is worse than c	$\tau=0$	

a is better than c	$\tau=0$	
a and c are indifferent	$\tau=0.541$	$(\alpha=0.40)$
a is worse than c	$\tau=0$	

a is better than c	$\tau=0$	
a and c are indifferent	$\tau=0$	$(\alpha=0.60)$
a is worse than c	$\tau=0$	

a is better than c	$\tau=0$	
a and c are indifferent	$\tau=0$	$(\alpha=0.80)$
a is worse than c	$\tau=0$	

a is better than d	$\tau=0$	
a and d are indifferent	$\tau=0.621$	$(\alpha=0.20)$
a is worse than d	$\tau=1$	

a is better than d	$\tau=0$	
a and d are indifferent	$\tau=0$	$(\alpha=0.40)$
a is worse than d	$\tau=0.69$	

a is better than d	$\tau=0$	
a and d are indifferent	$\tau=0$	$(\alpha=0.60)$
a is worse than d	$\tau=0$	

a is better than d	$\tau=0$	
a and d are indifferent	$\tau=0$	$(\alpha=0.80)$
a is worse than d	$\tau=0$	

b is better than c	$\tau=1$	
b and c are indifferent	$\tau=0$	$(\alpha=0.20)$
b is worse than c	$\tau=0.367$	

b is better than c	$\tau=1$	
b and c are indifferent	$\tau=0$	$(\alpha=0.40)$
b is worse than c	$\tau=0$	

b is better than c	$\tau=0$	
b and c are indifferent	$\tau=0$	$(\alpha=0.60)$
b is worse than c	$\tau=0$	
b is better than c	$\tau=0$	
b and c are indifferent	$\tau=0$	$(\alpha=0.80)$
b is worse than c	$\tau=0$	
b is better than d	$\tau=1$	
b and d are indifferent	$\tau=0$	$(\alpha=0.20)$
b is worse than d	$\tau=1$	
b is better than d	$\tau=0.174$	
b and d are indifferent	$\tau=0$	$(\alpha=0.40)$
b is worse than d	$\tau=0.028$	
b is better than d	$\tau=0$	
b and d are indifferent	$\tau=0$	$(\alpha=0.60)$
b is worse than d	$\tau=0$	
b is better than d	$\tau=0$	
b and d are indifferent	$\tau=0$	$(\alpha=0.80)$
b is worse than d	$\tau=0$	
c is better than d	$\tau=0$	
c and d are indifferent	$\tau=0.76$	$(\alpha=0.20)$
c is worse than d	$\tau=1$	
c is better than d	$\tau=0$	
c and d are indifferent	$\tau=0$	$(\alpha=0.40)$
c is worse than d	$\tau=0$	
c is better than d	$\tau=0$	
c and d are indifferent	$\tau=0$	$(\alpha=0.60)$
c is worse than d	$\tau=0$	
c is better than d	$\tau=0$	
c and d are indifferent	$\tau=0$	$(\alpha=0.80)$

c is worse than d $\tau=0$

By computing the ϕ^+ and ϕ^- indices by using the min operator (see equations from (7.25) to (7.28)), the following results are obtained:

<u>$\alpha=0.20$</u>

| $\phi^+(a)=0.659$ | $\phi^+(b)=0.992$ | $\phi^+(c)=0.586$ | $\phi^+(d)=0.977$ |
| $\phi^-(a)=0.802$ | $\phi^-(b)=0.839$ | $\phi^-(c)=0.81$ | $\phi^-(d)=0.604$ |

Then according to ϕ^+ the following ranking is obtained

b→   d→   a→   c

and according to ϕ^- the following ranking is obtained

d→   a→   c→   b

Therefore, the intersection gives the following final result:

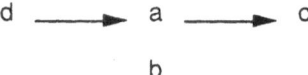

d ———▶ a ———▶ c

b

where action **b** results incomparable with all the other actions.

<u>$\alpha=0.40$</u>

| $\phi^+(a)=0.322$ | $\phi^+(b)=0.707$ | $\phi^+(c)=0.192$ | $\phi^+(d)=0.566$ |
| $\phi^-(a)=0.462$ | $\phi^-(b)=0.435$ | $\phi^-(c)=0.537$ | $\phi^-(d)=0.249$ |

Then according to ϕ^+ the following ranking is obtained

b→   d→   a→   c

and according to ϕ^- the following ranking is obtained

d→   b→   a→   c

Therefore, the intersection gives the following final result:

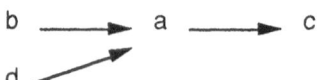

b ———▶ a ———▶ c

d

<u>$\alpha=0.60$</u>

| $\phi^+(a)=0.133$ | $\phi^+(b)=0.168$ | $\phi^+(c)=0.038$ | $\phi^+(d)=0.203$ |
| $\phi^-(a)=0.155$ | $\phi^-(b)=0.145$ | $\phi^-(c)=0.213$ | $\phi^-(d)=0$ |

Then according to ϕ^+ the following ranking is obtained d→   b→   a→   c

and according to ϕ^- the following ranking is obtained

d→   b→   a→   c

Therefore, the intersection gives the same result.

$\underline{\alpha=0.80}$

$\phi^+(a)=0.0173$	$\phi^+(b)=0.047$	$\phi^+(c)=0.004$	$\phi^+(d)=0.085$
$\phi^-(a)=0.048$	$\phi^-(b)=0.015$	$\phi^-(c)=0.051$	$\phi^-(d)=0$

Therefore, again the same result is obtained.

These results lead to the following comments. The pairwise linguistic evaluations offer useful complementary information to the final ranking. In fact, the former give indications of the relative credibility degree of preference, while the latter is only ordinal in nature. This is particular interesting for the evaluations on the indifference relations, since these are not considered in the ϕ^+ and ϕ^- indices. Moreover, one should note that in the ϕ^+ and ϕ^- indices the relationships among all the actions are considered, while the linguistic evaluations consider only a pair of actions. When high values of α are considered, it is quite difficult to arrive at a final pairwise linguistic evaluation. This is due to the presence of the threshold $\omega=0.5$ in the equation (7.21) and to the use of the min operator that does not allow any compensation between the values of the fuzzy relations and of the variable C. The presence of the threshold is also the reason where by means of the ϕ^+ and ϕ^- indices, it is still possible to rank actions for high values of α; however, one should note that in these situations only a few criteria with big intensities of preference could really play a role in the ranking process.

By looking to the previous results, one can see that for $\alpha=0.20$, in the final ranking **b** results incomparable with all the other actions. By taking into consideration the pairwise linguistic evaluations, one can see that for the pair **a, b** and **b, d** it is impossible to establish a preference or indifference relation since the fuzzy graph is equal to one in both directions. For the pair **b, c** in the light of the pairwise linguistic evaluation, the relation of incomparability seems less justified. For $\alpha=0.40$, the only incomparability relation that still exists is between **d** and **b**; this result is corroborated by the linguistic evaluations. The only possibility of indifference is between **a** and **c**. For $\alpha=0.60$ and $\alpha=0.80$ the pairwise linguistic evaluations do not discriminate anymore. However, we can conclude that the four rankings are consistent enough, and it is possible to isolate a group of best actions (**d** and **b**) and a group of worst actions (**a** and **c**) for any value of α considered.

As an example of the use of an aggregation operator with different properties from the min operator, in the following we will apply the Zimmermann-Zysno γ-operator (see equation (5.24)).

By using the Zimmermann-Zysno γ-operator (with $\gamma=0.6$), for each pair of actions, the degrees of truth of a linguistic evaluation illustrated in Appendix 2 are obtained. By computing the ϕ^+ and ϕ^- indices by using the Zimmermann-Zysno γ-operator, the following results are obtained:

$\underline{\alpha=0.20}$

$\phi^+(a)=0.826$ $\phi^+(b)=1.163$ $\phi^+(c)=0.742$ $\phi^+(d)=1.044$

$\phi^-(a)=0.919$ $\phi^-(b)=1.027$ $\phi^-(c)=1.019$ $\phi^-(d)=0.727$

Then according to ϕ^+ the following ranking is obtained

b$\rightarrow$ d$\rightarrow$ a$\rightarrow$ c

and according to ϕ^- the following ranking is obtained

d$\rightarrow$ a$\rightarrow$ c$\rightarrow$ b

Therefore, the intersection gives the following final result:

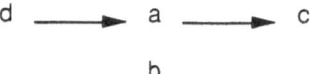

$\underline{\alpha=0.40}$

$\phi^+(a)=0.583$ $\phi^+(b)=0.843$ $\phi^+(c)=0.419$ $\phi^+(d)=0.758$

$\phi^-(a)=0.653$ $\phi^-(b)=0.622$ $\phi^-(c)=0.762$ $\phi^-(d)=0.489$

Then according to ϕ^+ the following ranking is obtained

b$\rightarrow$ d$\rightarrow$ a$\rightarrow$ c

and according to ϕ^- the following ranking is obtained

d$\rightarrow$ b$\rightarrow$ a$\rightarrow$ c

Therefore, the intersection gives the following final result:

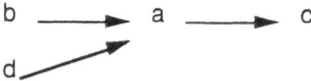

$\underline{\alpha=0.60}$

$\phi^+(a)=0.262$ $\phi^+(b)=0.368$ $\phi^+(c)=0.075$ $\phi^+(d)=0.51$

$\phi^-(a)=0.446$ $\phi^-(b)=0.364$ $\phi^-(c)=0.519$ $\phi^-(d)=0$

Then according to ϕ^+ the following ranking is obtained

d→   b→   a→   c

and according to ϕ^- the following ranking is obtained

d→   b→   a→   c

Therefore, the intersection gives the same result.

α=0.80

$\phi^+(a)=0.115$ $\phi^+(b)=0.174$ $\phi^+(c)=0.032$ $\phi^+(d)=0.243$

$\phi^-(a)=0.204$ $\phi^-(b)=0.109$ $\phi^-(c)=0.257$ $\phi^-(d)=0$

Again the same ranking is obtained.

When high values of α are considered, also in this case, it is quite difficult to arrive at a final pairwise linguistic evaluation; however by means of the Zimmermann-Zysno γ-operator, a higher degree of discrimination is possible. This is due to the fact that the γ-operator allows a certain degree of compensation between the values of the fuzzy relations and of the variable C. On the other hand, the use of the γ-operator does not seem to have any influence on the final rankings.

9.3 Intensity of Preference in the NAIADE Method: Some Particular Cases

In this section, we will illustrate some cases which show the role of the concept of intensity of preference in the NAIADE method. In particular, some examples in which results different from traditional efficiency analysis are obtained, will be presented. For the sake of simplicity, only crisp numbers will be used; the evaluations are supposed to be in a 0 - 100 interval scale.

For each fuzzy relation the following values of the parameter c are used:

$\mu_{\gg}$ c= 400

$\mu_>$ c= 372.6

$\mu_\cong$ c= 0.0693

$\mu_=$ c= 0.0277

For all the aggregations, the value α=0.40 is used.

Let us start with the example illustrated in Table 9.2. This is a clear case of dominance of action **a** over action **b**.

	Alternatives	
Criteria	a	b
g_1	100	30
g_2	100	40

Table 9.2 Evaluation Matrix of a Dominance Case

By using NAIADE the following pairwise fuzzy relations are obtained:

g_1

$\mu_{>>}(a, b)=0.863$
$\mu_{>}(a, b)=0.924$
$\mu_{\cong}(a, b)=0.0078$
$\mu_{=}(a, b)\cong 0$

g_2

$\mu_{>>}(a, b)=0.821$
$\mu_{>}(a, b)=0.9$
$\mu_{\cong}(a, b)=0.015$
$\mu_{=}(a, b)\cong 0$

Then, by applying the min operator, the following degrees of truth of a linguistic evaluation are obtained.

a is better than b	$\tau=1$
a and b are indifferent	$\tau=0$
a is worse than b	$\tau=0$

By applying the Zimmermann-Zysno γ-operator, we get the following degrees of truth of a linguistic evaluation.

a is better than b	$\tau=1$
a and b are indifferent	$\tau=0$
a is worse than b	$\tau=0$

In Table 9.3, another case of dominance is illustrated; traditional efficiency analysis would conclude that action **a** is better than action **b**, however one has to note that the intensity of preference of **a** over **b** could not be big enough to justify such a conclusion.

	Alternatives	
Criteria	a	b
g_1	100	95
g_2	100	95

Table 9.3 Evaluation Matrix of a Dominance Case

The following pairwise fuzzy relations are obtained:

g_1

g_2

$\mu_{\gg}(a, b)=0.003$

$\mu_{>}(a, b)=0.058$

$\mu_{\cong}(a, b)=0.707$

$\mu_{=}(a, b)=0.5$

$\mu_{\gg}(a, b)=0.003$

$\mu_{>}(a, b)=0.058$

$\mu_{\cong}(a, b)=0.707$

$\mu_{=}(a, b)=0.5$

By applying the min operator, the following degrees of truth of a linguistic evaluation are obtained.

a is better than b $\tau=0$

a and b are indifferent $\tau=1$

a is worse than b $\tau=0$

By applying the Zimmermann-Zysno γ-operator, the following degrees of truth of a linguistic evaluation are obtained.

a is better than b $\tau=0$

a and b are indifferent $\tau=1$

a is worse than b $\tau=0$

Now, we will take into consideration a case where both actions are efficient. As one can note in Table 9.4, the intensity of preference of **a** over **b** on the first criterion is exactly the same as the intensity of preference of **b** over **a** on the second criterion, thus to compare the two actions is quite difficult.

	Alternatives	
Criteria	a	b
g_1	100	30
g_2	30	100

Table 9.4 Evaluation Matrix of Two Efficient Actions

By applying NAIADE the following results are obtained:

g_1 g_2

$\mu_{>>}(a, b)=0.863$ $\mu_{\cong}(a, b)=0.0078$
$\mu_{>}(a, b)=0.924$ $\mu_{=}(a, b)\cong0$
$\mu_{\cong}(a, b)=0.0078$ $\mu_{<}(a, b)=0.924$
$\mu_{=}(a, b)\cong0$ $\mu_{<<}(a, b)=0.863$

By using the min operator, we get the following degrees of truth of a linguistic evaluation.

a is better than b $\tau=0.757$
a and b are indifferent $\tau=0$
a is worse than b $\tau=0.757$

Then by applying the Zimmermann-Zysno γ-operator, the following degrees of truth of a linguistic evaluation are obtained.

a is better than b $\tau=0.937$
a and b are indifferent $\tau=0$
a is worse than b $\tau=0.937$

Finally, let us take into consideration the case illustrated in Table 9.5; again actions **a** and **b** are both efficient, however now the intensity of preference of **b** over **a** is much bigger than the intensity of preference of **a** over **b**.

	Alternatives	
Criteria	a	b
g_1	100	95
g_2	30	100

Table 9.5 Evaluation Matrix of Two Efficient Actions

The following pairwise fuzzy relations are obtained:

g_1 g_2

$\mu_{>>}(a, b)=0.0039$ $\mu_{\cong}(a, b)=0.0078$
$\mu_{>}(a, b)=0.058$ $\mu_{=}(a, b)\cong 0$
$\mu_{\cong}(a, b)=0.707$ $\mu_{<}(a, b)=0.924$
$\mu_{=}(a, b)\cong 0.5$ $\mu_{<<}(a, b)=0.863$

By applying the min operator, the following degrees of truth of a linguistic evaluation are obtained.

a is better than b $\tau=0$
a and b are indifferent $\tau=0.464$
a is worse than b $\tau=0.759$

By applying the Zimmermann-Zysno γ-operator, the following degrees of truth of a linguistic evaluation are obtained.

a is better than b $\tau=0$
a and b are indifferent $\tau=0.841$
a is worse than b $\tau=0.937$

9.4   Concluding Remarks

Two critical factors in determining the results provided by the NAIADE method are the parameter α used in the equation (7.17) and the aggregation operator used for the approximate reasoning operations. Therefore, in applying NAIADE, the analysis of the robustness of the results obtained according to these two factors may be useful. In this Chapter an example of sensitivity analysis in the framework of NAIADE has been presented. In the light of the results obtained, the following conclusions can be drawn.

The pairwise linguistic evaluations offer useful complementary information to the final ranking. In fact, the former give indications of the relative credibility degree of preference, while the latter is only ordinal in nature. This is particularly interesting for the evaluations on the indifference relations, since these are not considered in the ϕ^+ and ϕ^- indices. Moreover, one should note that in the ϕ^+ and ϕ^- indices the relationships among all the actions are considered, while the linguistic evaluations consider only a pair of actions.

When high values of α are considered, it is quite difficult to arrive at a final pairwise linguistic evaluation. This is mainly due to the presence of the threshold $\omega=0.5$ in the equation (7.21); however by means of the Zimmermann-Zysno γ-operator, a higher degree of discrimination is possible. This is due to the fact that the γ-operator allows a certain degree of compensation between the values of the fuzzy relations and of the variable C, while the min operator is completely non-interactive. The presence of the threshold is also the reason why by means of the ϕ^+ and ϕ^- indices, it is still possible to rank actions for high values of α; however, one should note that in these situations only a few criteria with big intensities of preference could really play a role in the ranking process. In any case, we think that the use of high values of α may give some useful information above all when phenomena of rank reversal occur. In fact, in these cases the presence of rank reversals means that "discordant" criteria with big intensities of preference exist, thus information similar in spirit, to the one supplied by the veto threshold in ELECTRE methods could be obtained.

Finally, it should be noted that in our numerical example, the choice of the min operator or of the γ-operator only has little influence on the final rankings. This of course does not imply that this result is always true; furthermore, other operators exist that could supply different results, however as we have noted in section 5.6, the choice of a linguistic operator is a very difficult problem.

Since one of the main characteristics of NAIADE is to take into account the intensity of preference between criteria by means of fuzzy relations, some simple illustrative cases have also been presented. In some cases, results different from simple efficiency analysis are obtained, and overall the results are quite plausible.

APPENDIX 1

Pairwise Fuzzy Preference Relations

$$\left(\begin{array}{l} \mu_{>>}(a, b)=1 \\ \mu_{>}(a, b)=1 \\ \mu_{\cong}(a, b)=1 \\ \mu_{=}(a, b)=1 \\ \mu_{<}(a, b)=1 \\ \mu_{<<}(a, b)=1 \end{array}\right.$$

$C(>)=0.587$
$C(>>)=0.604$
$C(\cong)=0.575$
$C(=)=1$
$C(<)=0.758$
$C(<<)=0.753$

$(\alpha=0)$

$$\left(\begin{array}{l} \mu_{>>}(a, b)=0.642 \\ \mu_{>}(a, b)=0.705 \\ \mu_{\cong}(a, b)=0.298 \\ \mu_{=}(a, b)=0 \\ \mu_{<}(a, b)=0.75 \\ \mu_{<<}(a, b)=0.736 \end{array}\right.$$

$C(>)=0.587$
$C(>>)=0.604$
$C(\cong)=0.575$
$C(=)=1$
$C(<)=0.758$
$C(<<)=0.753$

$(\alpha=0.20)$

$$\left(\begin{array}{l} \mu_{>>}(a, b)=0.309 \\ \mu_{>}(a, b)=0.411 \\ \mu_{\cong}(a, b)=0 \\ \mu_{=}(a, b)=0 \\ \mu_{<}(a, b)=0.5 \\ \mu_{<<}(a, b)=0.47 \end{array}\right.$$

$C(>)=0.815$
$C(>>)=0.604$
$C(\cong)=1$
$C(=)=1$
$C(<)=0.758$
$C(<<)=0.753$

$(\alpha=0.40)$

$$\left(\begin{array}{l} \mu_{>>}(a, b)=0.114 \\ \mu_{>}(a, b)=0.116 \\ \mu_{\cong}(a, b)=0 \\ \mu_{=}(a, b)=0 \\ \mu_{<}(a, b)=0.25 \\ \mu_{<<}(a, b)=0.27 \end{array}\right.$$

$C(>)=0.815$
$C(>>)=0.854$
$C(\cong)=1$
$C(=)=1$
$C(<)=1$
$C(<<)=1$

$(\alpha=0.60)$

$$\left(\begin{array}{l} \mu_{>>}(a, b)=0 \\ \mu_>(a, b)=0.30 \\ \mu_\cong(a, b)=0 \\ \mu_=(a, b)=0 \\ \mu_<(a, b)=0.1 \\ \mu_{<<}(a, b)=0.09 \end{array}\right.$$

$$\begin{array}{l} C(>)=1 \\ C(>>)=0.854 \\ C(\cong)=1 \\ C(=)=1 \\ C(<)=1 \\ C(<<)=1 \end{array}$$

($\alpha=0.80$)

$$\left(\begin{array}{l} \mu_{>>}(a, c)=1 \\ \mu_>(a, c)=1 \\ \mu_\cong(a, c)=1 \\ \mu_=(a, c)=1 \\ \mu_<(a, c)=1 \\ \mu_{<<}(a, c)=1 \end{array}\right.$$

$$\begin{array}{l} C(>)=0.848 \\ C(>>)=0.9 \\ C(\cong)=0.769 \\ C(=)=1 \\ C(<)=0.75 \\ C(<<)=0.753 \end{array}$$

($\alpha=0$)

$$\left(\begin{array}{l} \mu_{>>}(a, c)=0.52 \\ \mu_>(a, c)=0.545 \\ \mu_\cong(a, c)=0.89 \\ \mu_=(a, c)=0.8 \\ \mu_<(a, c)=0.33 \\ \mu_{<<}(a, c)=0.354 \end{array}\right.$$

$$\begin{array}{l} C(>)=0.848 \\ C(>>)=0.9 \\ C(\cong)=0.769 \\ C(=)=1 \\ C(<)=0.75 \\ C(<<)=0.753 \end{array}$$

($\alpha=0.20$)

$$\left(\begin{array}{l} \mu_{>>}(a, c)=0.272 \\ \mu_>(a, c)=0.302 \\ \mu_\cong(a, c)=0.722 \\ \mu_=(a, c)=0.6 \\ \mu_<(a, c)=0.076 \\ \mu_{<<}(a, c)=0.097 \end{array}\right.$$

$$\begin{array}{l} C(>)=0.848 \\ C(>>)=0.9 \\ C(\cong)=1 \\ C(=)=1 \\ C(<)=0.875 \\ C(<<)=0.753 \end{array}$$

($\alpha=0.40$)

$$\left(\begin{array}{l} \mu_{>>}(a, c)=0.121 \\ \mu_>(a, c)=0.15 \\ \mu_\cong(a, c)=0.481 \\ \mu_=(a, c)=0.4 \\ \mu_<(a, c)=0 \\ \mu_{<<}(a, c)=0 \end{array}\right.$$

$$\begin{array}{l} C(>)=0.848 \\ C(>>)=0.9 \\ C(\cong)=1 \\ C(=)=1 \\ C(<)=1 \\ C(<<)=1 \end{array}$$

($\alpha=0.60$)

$$\left(\begin{array}{l} \mu_{>>}(a,c)=0.02 \\ \mu_{>}(a,c)=0.047 \\ \mu_{\cong}(a,c)=0.24 \\ \mu_{=}(a,c)=0.2 \\ \mu_{<}(a,c)=0 \\ \mu_{<<}(a,c)=0 \end{array}\right.$$

$$\begin{array}{l} C(>)=0.848 \\ C(>>)=0.9 \\ C(\cong)=1 \\ C(=)=1 \\ C(<)=1 \\ C(<<)=1 \end{array}$$

$(\alpha=0.80)$

$$\left(\begin{array}{l} \mu_{>>}(a,d)=1 \\ \mu_{>}(a,d)=1 \\ \mu_{\cong}(a,d)=1 \\ \mu_{=}(a,d)=1 \\ \mu_{<}(a,d)=1 \\ \mu_{<<}(a,d)=1 \end{array}\right.$$

$$\begin{array}{l} C(>)=0.753 \\ C(>>)=0.758 \\ C(\cong)=0.362 \\ C(=)=1 \\ C(<)=0.658 \\ C(<<)=0.601 \end{array}$$

$(\alpha=0)$

$$\left(\begin{array}{l} \mu_{>>}(a,d)=0.347 \\ \mu_{>}(a,d)=0.333 \\ \mu_{\cong}(a,d)=0.811 \\ \mu_{=}(a,d)=0.571 \\ \mu_{<}(a,d)=0.736 \\ \mu_{<<}(a,d)=0.708 \end{array}\right.$$

$$\begin{array}{l} C(>)=0.753 \\ C(>>)=0.758 \\ C(\cong)=0.362 \\ C(=)=1 \\ C(<)=0.658 \\ C(<<)=0.601 \end{array}$$

$(\alpha=0.20)$

$$\left(\begin{array}{l} \mu_{>>}(a,d)=0.09 \\ \mu_{>}(a,d)=0.125 \\ \mu_{\cong}(a,d)=0.447 \\ \mu_{=}(a,d)=0.333 \\ \mu_{<}(a,d)=0.473 \\ \mu_{<<}(a,d)=0.416 \end{array}\right.$$

$$\begin{array}{l} C(>)=0.753 \\ C(>>)=0.758 \\ C(\cong)=1 \\ C(=)=1 \\ C(<)=0.658 \\ C(<<)=0.601 \end{array}$$

$(\alpha=0.40)$

$$\left(\begin{array}{l} \mu_{>>}(a,d)=0 \\ \mu_{>}(a,d)=0 \\ \mu_{\cong}(a,d)=0.229 \\ \mu_{=}(a,d)=0.181 \\ \mu_{<}(a,d)=0.21 \\ \mu_{<<}(a,d)=0.163 \end{array}\right.$$

$$\begin{array}{l} C(>)=1 \\ C(>>)=1 \\ C(\cong)=1 \\ C(=)=1 \\ C(<)=0.9 \\ C(<<)=0.848 \end{array}$$

$(\alpha=0.60)$

$$
\left(
\begin{array}{l}
\mu_{>>}(a, d)=0 \\
\mu_{>}(a, d)=0 \\
\mu_{\cong}(a, d)=0.093 \\
\mu_{=}(a, d)=0.076 \\
\mu_{<}(a, d)=0.0625 \\
\mu_{<<}(a, d)=0.025
\end{array}
\right.
\qquad
\begin{array}{l}
C(>)=1 \\
C(>>)=1 \\
C(\cong)=1 \\
C(=)=1 \\
C(<)=0.9 \\
C(<<)=0.848
\end{array}
\qquad (\alpha=0.80)
$$

$$
\left(
\begin{array}{l}
\mu_{>>}(b, c)=1 \\
\mu_{>}(b, c)=1 \\
\mu_{\cong}(b, c)=1 \\
\mu_{=}(b, c)=1 \\
\mu_{<}(b, c)=1 \\
\mu_{<<}(b, c)=1
\end{array}
\right.
\qquad
\begin{array}{l}
C(>)=0.339 \\
C(>>)=0.474 \\
C(\cong)=0.575 \\
C(=)=1 \\
C(<)=0.854 \\
C(<<)=0.815
\end{array}
\qquad (\alpha=0)
$$

$$
\left(
\begin{array}{l}
\mu_{>>}(b, c)=0.838 \\
\mu_{>}(b, c)=0.863 \\
\mu_{\cong}(b, c)=0.298 \\
\mu_{=}(b, c)=0 \\
\mu_{<}(b, c)=0.523 \\
\mu_{<<}(b, c)=0.495
\end{array}
\right.
\qquad
\begin{array}{l}
C(>)=0.339 \\
C(>>)=0.474 \\
C(\cong)=0.575 \\
C(=)=1 \\
C(<)=0.854 \\
C(<<)=0.815
\end{array}
\qquad (\alpha=0.20)
$$

$$
\left(
\begin{array}{l}
\mu_{>>}(b, c)=0.520 \\
\mu_{>}(b, c)=0.655 \\
\mu_{\cong}(b, c)=0 \\
\mu_{=}(b, c)=0 \\
\mu_{<}(b, c)=0.277 \\
\mu_{<<}(b, c)=0.245
\end{array}
\right.
\qquad
\begin{array}{l}
C(>)=0.568 \\
C(>>)=0.474 \\
C(\cong)=1 \\
C(=)=1 \\
C(<)=0.854 \\
C(<<)=0.815
\end{array}
\qquad (\alpha=0.40)
$$

$$
\left(
\begin{array}{l}
\mu_{>>}(b, c)=0.166 \\
\mu_{>}(b, c)=0.270 \\
\mu_{\cong}(b, c)=0 \\
\mu_{=}(b, c)=0 \\
\mu_{<}(b, c)=0.126 \\
\mu_{<<}(b, c)=0.095
\end{array}
\right.
\qquad
\begin{array}{l}
C(>)=0.815 \\
C(>>)=0.854 \\
C(\cong)=1 \\
C(=)=1 \\
C(<)=0.854 \\
C(<<)=0.815
\end{array}
\qquad (\alpha=0.60)
$$

$$
\left(
\begin{array}{l}
\mu_{>>}(b, c)=0 \\
\mu_{>}(b, c)=0.0441 \\
\mu_{\cong}(b, c)=0 \\
\mu_{=}(b, c)=0 \\
\mu_{<}(b, c)=0.024 \\
\mu_{<<}(b, c)=0
\end{array}
\right.
\qquad
\begin{array}{l}
C(>)=1 \\
C(>>)=0.854 \\
C(\cong)=1 \\
C(=)=1 \\
C(<)=0.854 \\
C(<<)=1
\end{array}
\qquad (\alpha=0.80)
$$

$$
\left(
\begin{array}{l}
\mu_{>>}(b, d)=1 \\
\mu_{>}(b, d)=1 \\
\mu_{\cong}(b, d)=1 \\
\mu_{=}(b, d)=1 \\
\mu_{<}(b, d)=1 \\
\mu_{<<}(b, d)=1
\end{array}
\right.
\qquad
\begin{array}{l}
C(>)=0.505 \\
C(>>)=0.515 \\
C(\cong)=0.381 \\
C(=)=1 \\
C(<)=0.604 \\
C(<<)=0.587
\end{array}
\qquad (\alpha=0)
$$

$$
\left(
\begin{array}{l}
\mu_{>>}(b, d)=0.615 \\
\mu_{>}(b, d)=0.666 \\
\mu_{\cong}(b, d)=0.5 \\
\mu_{=}(b, d)=0 \\
\mu_{<}(b, d)=0.705 \\
\mu_{<<}(b, d)=0.642
\end{array}
\right.
\qquad
\begin{array}{l}
C(>)=0.505 \\
C(>>)=0.515 \\
C(\cong)=0.381 \\
C(=)=1 \\
C(<)=0.604 \\
C(<<)=0.587
\end{array}
\qquad (\alpha=0.20)
$$

$$
\left(
\begin{array}{l}
\mu_{>>}(b, d)=0.23 \\
\mu_{>}(b, d)=0.333 \\
\mu_{\cong}(b, d)=0 \\
\mu_{=}(b, d)=0 \\
\mu_{<}(b, d)=0.411 \\
\mu_{<<}(b, d)=0.309
\end{array}
\right.
\qquad
\begin{array}{l}
C(>)=0.505 \\
C(>>)=0.515 \\
C(\cong)=1 \\
C(=)=1 \\
C(<)=0.604 \\
C(<<)\approx0.815
\end{array}
\qquad (\alpha=0.40)
$$

$$
\left(
\begin{array}{l}
\mu_{>>}(b, d)=0 \\
\mu_{>}(b, d)=0 \\
\mu_{\cong}(b, d)=0 \\
\mu_{=}(b, d)=0 \\
\mu_{<}(b, d)=0.166 \\
\mu_{<<}(b, d)=0.114
\end{array}
\right.
\qquad
\begin{array}{l}
C(>)=1 \\
C(>>)=1 \\
C(\cong)=1 \\
C(=)=1 \\
C(<)=0.854 \\
C(<<)=0.815
\end{array}
\qquad (\alpha=0.60)
$$

$$
\left\{
\begin{array}{ll}
\mu_{>>}(b, d)=0 & C(>)=1 \\
\mu_{>}(b, d)=0 & C(>>)=1 \\
\mu_{\cong}(b, d)=0 & C(\cong)=1 \\
\mu_{=}(b, d)=0 & C(=)=1 \\
\mu_{<}(b, d)=0.03 & C(<)=0.854 \\
\mu_{<<}(b, d)=0 & C(<<)=1
\end{array}
\right. \qquad (\alpha=0.80)
$$

$$
\left\{
\begin{array}{ll}
\mu_{>>}(c, d)=1 & C(>)=0.753 \\
\mu_{>}(c, d)=1 & C(>>)=0.758 \\
\mu_{\cong}(c, d)=1 & C(\cong)=0.575 \\
\mu_{=}(c, d)=1 & C(=)=1 \\
\mu_{<}(c, d)=1 & C(<)=0.75 \\
\mu_{<<}(c, d)=1 & C(<<)=0.772
\end{array}
\right. \qquad (\alpha=0)
$$

$$
\left\{
\begin{array}{ll}
\mu_{>>}(c, d)=0.347 & C(>)=0.753 \\
\mu_{>}(c, d)=0.4 & C(>>)=0.758 \\
\mu_{\cong}(c, d)=0.829 & C(\cong)=0.575 \\
\mu_{=}(c, d)=0.571 & C(=)=1 \\
\mu_{<}(c, d)=0.733 & C(<)=0.75 \\
\mu_{<<}(c, d)=0.699 & C(<<)=0.772
\end{array}
\right. \qquad (\alpha=0.20)
$$

$$
\left\{
\begin{array}{ll}
\mu_{>>}(c, d)=0.09 & C(>)=0.753 \\
\mu_{>}(c, d)=0.142 & C(>>)=0.758 \\
\mu_{\cong}(c, d)=0.346 & C(\cong)=1 \\
\mu_{=}(c, d)=0.338 & C(=)=1 \\
\mu_{<}(c, d)=0.466 & C(<)=0.75 \\
\mu_{<<}(c, d)=0.408 & C(<<)=1
\end{array}
\right. \qquad (\alpha=0.40)
$$

$$
\left\{
\begin{array}{ll}
\mu_{>>}(c, d)=0 & C(>)=1 \\
\mu_{>}(c, d)=0 & C(>>)=1 \\
\mu_{\cong}(c, d)=0.261 & C(\cong)=1 \\
\mu_{=}(c, d)=0.181 & C(=)=1 \\
\mu_{<}(c, d)=0.235 & C(<)=1 \\
\mu_{<<}(c, d)=0.213 & C(<<)=1
\end{array}
\right. \qquad (\alpha=0.60)
$$

$$\left\{\begin{array}{ll} \mu_{>>}(c, d)=0 & C(>)=1 \\ \mu_{>}(c, d)=0 & C(>>)=1 \\ \mu_{\cong}(c, d)=0.098 & C(\cong)=1 \\ \mu_{=}(c, d)=0.076 & C(=)=1 \\ \mu_{<}(c, d)=0.095 & C(<)=1 \\ \mu_{<<}(c, d)=0.08 & C(<<)=1 \end{array}\right. \qquad (\alpha=0.80)$$

APPENDIX 2

Degrees of Truth Obtained by Means of the γ-Operator

a is better than b	$\tau=1$	
a and b are indifferent	$\tau=0$	$(\alpha=0.20)$
a is worse than b	$\tau=1$	
a is better than b	$\tau=0.724$	
a and b are indifferent	$\tau=0$	$(\alpha=0.40)$
a is worse than b	$\tau=1$	
a is better than b	$\tau=0$	
a and b are indifferent	$\tau=0$	$(\alpha=0.60)$
a is worse than b	$\tau=0.281$	
a is better than b	$\tau=0$	
a and b are indifferent	$\tau=0$	$(\alpha=0.80)$
a is worse than b	$\tau=0$	
a is better than c	$\tau=1$	
a and c are indifferent	$\tau=1$	$(\alpha=0.20)$
a is worse than c	$\tau=0.647$	
a is better than c	$\tau=0.404$	
a and c are indifferent	$\tau=1$	$(\alpha=0.40)$
a is worse than c	$\tau=0$	
a is better than c	$\tau=0$	
a and c are indifferent	$\tau=0.734$	$(\alpha=0.60)$
a is worse than c	$\tau=0$	
a is better than c	$\tau=0$	
a and c are indifferent	$\tau=0.154$	$(\alpha=0.80)$
a is worse than c	$\tau=0$	
a is better than d	$\tau=0$	

| a and d are indifferent | $\tau=1$ | ($\alpha=0.20$) |
| a is worse than d | $\tau=1$ | |

a is better than d	$\tau=0$	
a and d are indifferent	$\tau=0.617$	($\alpha=0.40$)
a is worse than d	$\tau=1$	

a is better than d	$\tau=0$	
a and d are indifferent	$\tau=0.101$	($\alpha=0.60$)
a is worse than d	$\tau=0.058$	

a is better than d	$\tau=0$	
a and d are indifferent	$\tau=0$	($\alpha=0.80$)
a is worse than d	$\tau=0$	

b is better than c	$\tau=1$	
b and c are indifferent	$\tau=0$	($\alpha=0.20$)
b is worse than c	$\tau=1$	

b is better than c	$\tau=1$	
b and c are indifferent	$\tau=0$	($\alpha=0.40$)
b is worse than c	$\tau=0$	

b is better than c	$\tau=0.184$	
b and c are indifferent	$\tau=0$	($\alpha=0.60$)
b is worse than c	$\tau=0$	

b is better than c	$\tau=0$	
b and c are indifferent	$\tau=0$	($\alpha=0.80$)
b is worse than c	$\tau=0$	

b is better than d	$\tau=1$	
b and d are indifferent	$\tau=0$	($\alpha=0.20$)
b is worse than d	$\tau=1$	

b is better than d	$\tau=0.641$	
b and d are indifferent	$\tau=0$	($\alpha=0.40$)
b is worse than d	$\tau=0.724$	

192

b is better than d	$\tau=0$	
b and d are indifferent	$\tau=0$	$(\alpha=0.60)$
b is worse than d	$\tau=0$	
b is better than d	$\tau=0$	
b and d are indifferent	$\tau=0$	$(\alpha=0.80)$
b is worse than d	$\tau=0$	
c is better than d	$\tau=0.737$	
c and d are indifferent	$\tau=1$	$(\alpha=0.20)$
c is worse than d	$\tau=1$	
c is better than d	$\tau=0$	
c and d are indifferent	$\tau=0.504$	$(\alpha=0.40)$
c is worse than d	$\tau=0.81$	
c is better than d	$\tau=0$	
c and d are indifferent	$\tau=0.151$	$(\alpha=0.60)$
c is worse than d	$\tau=0.168$	
c is better than d	$\tau=0$	
c and d are indifferent	$\tau=0$	$(\alpha=0.80)$
c is worse than d	$\tau=0$	

PART C

APPLICATION TO A REAL-WORLD ENVIRONMENTAL MANAGEMENT PROBLEM

CHAPTER 10

THE RIVER PO BASIN ENVIRONMENTAL POLICY. BACKGROUND INFORMATION AND ANALYTICAL FRAMEWORK

10.1 Introduction

In this part of the study, we will apply the methods developed earlier to a real-world environmental management problem in the river Po basin. Chapter 10 will describe the main characteristics of the case study; in Chapter 11 the decision analysis will be carried out.

The process of eutrophication of the coastal waters of Emilia-Romagna, because of the intensity of the algal blooms and the extent of the area involved, forms a unique case in the Mediterranean. If one considers that in supplying nutrient salts in the surface water layer where photosynthesis may take place, the main sources are represented by contributions from land, it is not surprising that the coastal band of this part of the Mediterranean, characterised by low bottoms and into which various important watercourses flow, has been known for some time for its high algal productivity which recently has reached explosive points.

The consequences of these blooms are well known. As a direct effect when the abnormal algal biomass decomposes it takes oxygen from the waters until they become completely anoxic. This condition may be triggered when the physical process of reaeration and therefore the passage of new oxygen from the atmosphere to the water does not occur sufficiently quickly to compensate for the requirements of the algal biomass which is decomposing in the deep layers. Anoxic conditions are thus created which are incompatible with the survival of animal life, especially those species which live on the sea bottom [Marchetti, 1984].

The abnormal development of algae and the cited beachings of large quantities of dead animals, the development of evil-smelling gases (especially hydrogen sulphide) which form in the bottom waters and which are released when the beached material decomposes, as well as the specific damage to fish, may cause conditions which are incompatible with the performance of other activities, typically swimming and local tourism, posing serious problems not only of appearance but also of hygiene and health.

From the hydrodynamic point of view the situation was first examined only in a sample area between Cesenatico and the mouth of the river Savio. This examination arrived at the following conclusion: "the area studied is characterised by a clear separation between the surface layer and the deep layer. The former normally contains 20 - 40% water of river origin and can contain up to 60% during Po floods. This stratification reduces and disappears under the coast in a 2 km band. It also disappears generally in stormy sea conditions. The circulation of the surface layer waters is closely dependent on the contributions from the Po, whose influence seems to be greatest when there are winds from the north at the same time as Po floods" [Marchetti, 1984].

The importance of these preliminary findings, which show the important role of the Po, pose the need for a check of the hydrodynamic situation over longer times and over wider areas. These findings lead to the conclusion that the chemical characteristics of the area are strongly determined as to time by the Po contributions which condition the general physiognomy and by those of the watercourses and other coastal immissions which are responsible for quite considerable pollution which in general is rather localised. From the data described one can see that the contributions of the Po hinterland and coast also affect the bottom layer for which, on the basis of current measurements, one could suppose a greater isolation. The dependence of bottom water characteristics on phenomena which occur at the surface is particularly clear when there are large algal bloom which are reflected in a fall in the oxygen reserve.

We can here briefly recall that the possibilities of reducing pollution loads are based on

(a)   *prevention ("upstream") of the formation of loads.* The formation of (e.g. phosphorus) loads which reach the waters can be prevented by "upstream" action. In the case of industry this may be by alternative technologies and processes, the rationalisation of some agricultural processes, modifications in the operation of animal rearing activities and the introduction of compounds to replace tripolyphosphates in detergents.

(b)   *Reduction (" downstream") of loads before they are emitted in the waters* which can be attained, mainly, by biological and chemical purification of human, agricultural and industrial waste.

(c)   *Dispersion of loads in the sea* with systems of pipes or by "dumping", to avoid the reaching of phosphorus concentrations which locally favour the onset of blooms.

It is clear that the upstream action is mainly connected to the cleaning up of the river Po basin. This action is quite complex mainly due to four factors [Premazzi et al., 1990]:

(1) *the size of the basin* (about a quarter of the area of Italy);

(2) *the size of the population* (about a third of the Country's total population, including the two largest urban and industrial concentrations);

(3) *the gross product* (about 40% of the GNP. The Po valley hosts about 34% of the Italian industry, which employs about 51% of Italian man power, and has one of the most productive agricultural systems including an area of about 49% of the area of the entire river Po basin; this area also supports about 50% of all the livestock reared in Italy);

(4) *the variety of the use of water sources* (drinking water, irrigation, electricity production).

The river Po basin is delimited by the Alps, which fix, from West to North, the boundaries with France, Switzerland and Austria. The southern limit is set by the northern Apennines, while towards the East the catchment faces the Adriatic Sea (see Figure 10.1). The basin runs from West to East, the most important Italian lakes Maggiore, Como, Iseo and Garda, lie near the transition between the two main regions of the basin, the mountain arc and the lowland plain. The cities of Turin and Milan lie on the plain, with about 1.3 and 1.8 million inhabitants respectively. The river Po length is of 677 km and the river Po source is on the Monviso glacier (3841 m).

10.2 Pollution Generation Factors

Regarding the water quality of the river Po tributaries organic pollution indicators allow us to form three groups of watercourse [Ministero dell'Ambiente, 1991]. The first comprises the Lambro, Parma and Crostolo, with high values of BOD5 (biochemical oxygen demand), COD (chemical oxygen demand), ammonia and phosphates. These waters allow the frequent isolation of pathogenic micro-organisms. The surfactants also reach particularly high levels, which exceed the risk threshold for aquatic life.

The second group includes the Ticino, Nure, Trebbia and Adda where BOD5, COD, ammonia, phosphates and detergents are on average 5 - 10 times less than those of the preceding groups. For the majority of the parameters considered the third group, containing the Arda, Ongina, Tara and Oglio, has levels in between those of the groups considered above. In most of the

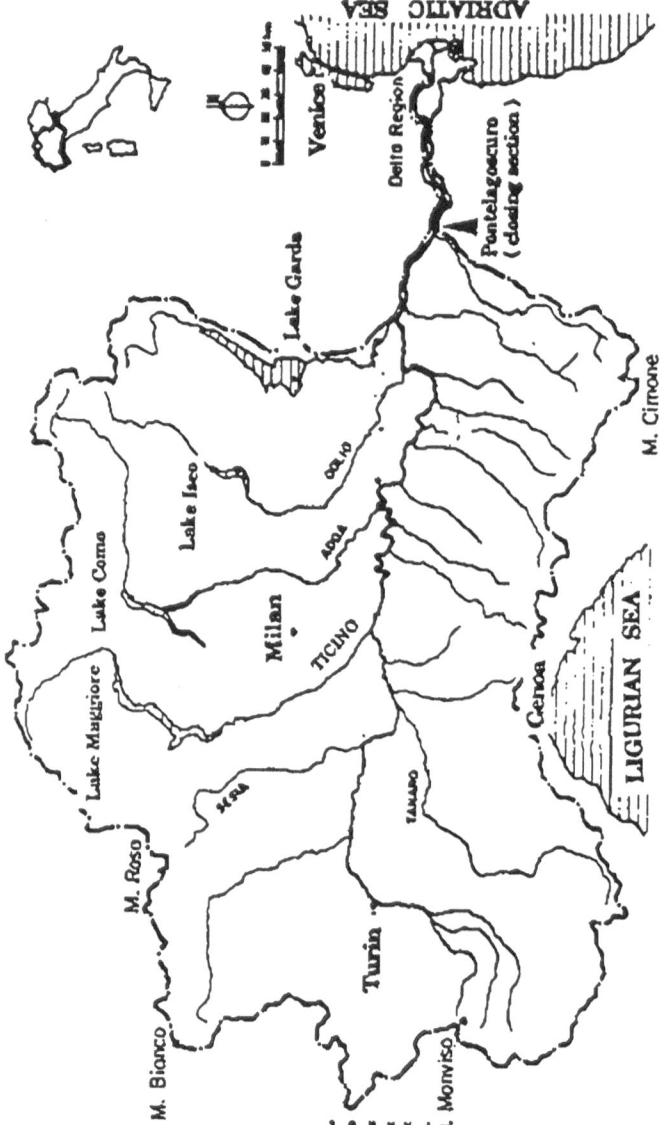

Figure 10.1   The River Po Basin

tributaries of the second and third group however the levels of faecal micro-organisms and the frequent isolation of pathogenic micro-organisms indicate a potential danger of infection.

For the river Po, for swimming with direct contact the judgement is negative for all the stations considered. For drinking use, the main cause of the negative judgement is microbial pollution. If one hypothetically neglects this factor, however, the judgement remains negative because of the high frequency of over-the-limit values for BOD5, ammonia, mercury and especially for phenols which reach high values for the whole stretch considered. For aquatic life analysis of the data does not change the conclusions given for other purposes of use. In fact in most stations the values exceed nearly all the parameters. The analytical values considered lead to a very negative overall judgement on the general state of the Po waters, and in a more definite way on the middle-lower stretch faced by the area of Cremona. However, while the condition of the Po river waters remains worrying in reference to aquatic life and other uses, at present there are no side effects for its use in irrigation [Ministero dell' Agricoltura, 1990].

The three polluting components - civil, industrial and agricultural - make river Po waters unsuitable for swimming, the development of a normal aquatic life and the supplying of water systems for human consumption. The identification of pollutant generation factors is tackled more thoroughly in the following sectors, indicating for each the specific characteristics of the pollution produced:

- Civil anthropic activity
- Industrial activity
- Animal rearing activity
- Agricultural activity

In May 1989 the Italian Parliament passes the law 183 on soil protection which lays down the establishment of the basin Authority. This law introduces the hydrographic basin into the area of physical planning, thereby overcoming problems of fragmentation and separation previously caused by administrative procedures. The Conference has chosen to work through a general overall plan, called the MASTERPLAN [Ministero dell' Ambiente, 1991], which intends to be a planning and programming tool.

The basin of the Po river is constituted of an area delimited in the MASTERPLAN co-ordination phase as a function of the main aim of the "Plan for cleaning up, safeguarding and exploiting the hydrographic basin of the river

Po", intended as a planning and programming tool which can allow the identification of initiatives aimed both at recovering the environmental quality of the basin, and at satisfying the demand for environmental resources, with particular reference to water resources.

The area under examination includes the whole hydrographic basin of the river Po and the system of surface and underground watercourses forming part of the same basin, thus extending over a very wide area and being very diversified, both in terms of surface and of residential and productive sites which demand water resources, but also generate consumption and deterioration of water and environmental resources.

10.2.1 Civil Anthropic Activity

The area considered extends for an overall area of about 71055 km^2, 23.6% of the area of Italy. It includes 27 provinces and 3188 local authority districts divided as follows:

- 1209 communes of the whole Piedmont Region;
- 74 communes of the whole Aosta Valley Region;
- 1540 communes of the Lombardy Region of a total of 1546 communes of this region, excluding 6 communes of the Province of Mantua;
- 221 communes belonging to the Emilia-Romagna region. With respect to a total of 341 communes of this region, the area of study included the provincial districts of Piacenza, Parma, Reggio Emilia, Modena and Ferrara and 8 of the 60 communes which form the province of Bologna.
- 64 communes of the Trent province out of 339 communes of the Trentino Alto Adige Region;
- 18 communes belonging to the Veneto Region, which has 582 communes altogether. Of these 18 communes 14 are from the Verona province and 4 from the Rovigo province;
- 62 communes belonging to the Ligurian Region, which has 235 communal districts. Of these 62 communes 29 are from the Savona province and 33 from the Genoa province.

The area considered overall, as has already been mentioned, represents less than 1/4 of the area of Italy but is characterised by very high area concentration indices, both in terms of environmental potential because of the residential and productive areas, and because of the use they make of the area. The area has about 15.800.000 inhabitants (1989 population), with a distribution at the level of the individual basin given in Table 10.1. Table 10.1 also reports the values of the surface areas of each basin and the density of inhabitation expressed in inhabitants/km^2.

BASIN	AREA Km^2	POPULATION	DENSITY Inhab/Km^2
01 MINCIO	3553	410930	116
02 OGLIO	6915	1382299	200
03 ADDA	8098	1832012	226
04 LAMBRO-SEVESO-OLONA	2952	4366672	1479
05 TICINO	4314	974240	226
06 AGOGNA	1763	324053	184
07 SESIA	3246	401093	124
08 DORA BALTEA	4085	264994	65
09 ORCO-MALONE-STURA DI LANZO	2231	326621	146
10 DORA RIPARIA	1332	1254243	942
11 SANGONE-CHISOLA-PELLICE-CHISONE	1700	317015	186
12 ALTO PO	2574	176886	69
13 RICCHIARDO BANNA	678	192219	284
14 MONFERRATO	944	158175	168
15 TANARO	8579	891147	104
16 SCRIVIA	1154	110831	96
17 CURONE-STRAFFORA-VERSA-TIDONE	1829	202326	111
18 NURE-TREBBIA	1756	166589	95
19 CHIAVENNA-ARDA-ONGINA	784	68945	88
20 TARO	2156	147001	68
21 PARMA	877	216018	246
22 ENZA-CROSTOLO	1486	276427	186
23 OLTREPO MANTOVANO	562	93235	166
24 SECCHIA	1967	315869	161
25 PANARO	2006	434169	216
26 BURANA-PO DI VOLANO	3112	434022	139
27 DELTA DEL PO	405	27528	68
	71057	15765559	222

Table 10.1   Sub-Basins of the River Po Hydrographic Basin
(From MASTERPLAN)

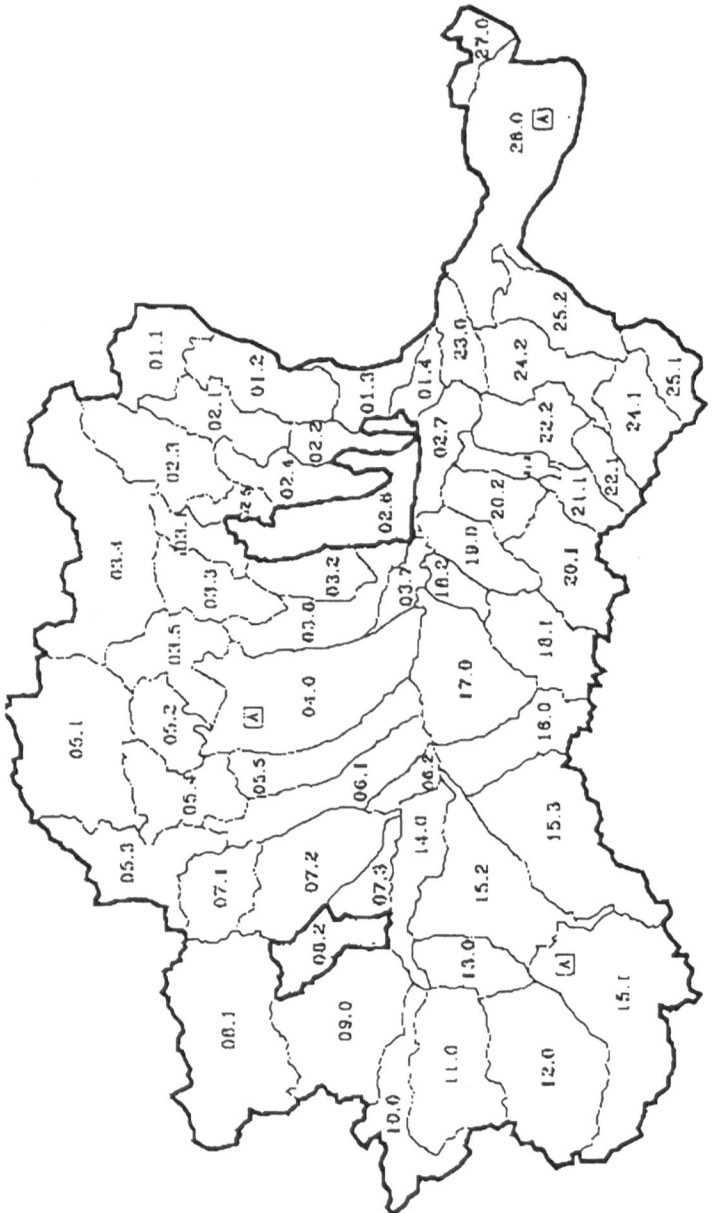

Figure 10.2   Geographic Location of the Various Basins

In Figure 10.2 the geographic location of the various basins is shown.

In the ambit of this anthropic activity three sectors of investigation can be identified concerning liquid discharges, production of solid urban waste and emissions in the atmosphere (domestic heating and traffic).

The parameters considered typical for this pollution source are: BOD5, Nitrogen, Phosphorus, COD, Surfactants and faecal Coliforms, Sulphur oxides, Nitrogen oxides, Carbon monoxide, Volatile Organic Compounds, Total Suspended Particles and Lead.

10.2.2   Industrial Activity

In the whole Po basin 2.831.000 people work in the productive activities examined (ISTAT 1981 datum). Most workers in the basin occupy ISTAT classes 31 (Industry of the construction of metal products), with 13.3%, 32 (Industry of the construction and installation of machines and mechanical materials), with 11% and 43 (textile industries) with 9.8%.

The basin which registers the most people working in the productive activities which are the subject of the investigation is the "Lambro-Seveso-Olona" basin where 30.3% of the workers of the Po basin live. The Lambro basin is followed by the basins of the Adda (10.6%), the Oglio (8.9%), the Dora Riparia (8.8%) and the Ticino (6.7%). All the other basins have percentage values lower than 5%.

10.2.3   Animal-Rearing Activity

The data for the number of head of cattle, pigs and sheep in the Communes of the Po basin can be used to determine the load generated by animal rearing. In the river Po basin more than 5 million pigs and 4 million cattle are reared, equal to 56% and 47% of the corresponding national values; the rearing of sheep is not important.

From the territorial distribution of animal rearing one can see the considerable weight of the Regions of Lombardy, Emilia-Romagna and Piedmont in both cattle and pig rearing. In fact Lombardy has about 2 million cattle and over 2.7 million pigs, Emilia-Romagna more than 850.000 cattle and 1.8 million pigs and finally Piedmont has almost 1.2 million cattle and 700.000 pigs. One should also remember that the cattle rearing situation has changed greatly over the years thanks also to the rapid evolution of production processes. The technological progress following the search for accelerated growth of cattle, forcing the forage-growth ratios, has revealed itself on the one hand in the intensification of monocrops and on the other in the concentration of cattle in stock farms without earth with ever greater problems of environmental impact because of the need to deal with the enormous quantities of excrement produced.

10.2.4   Agricultural Activity

In recent years there has been a considerable evolution of agriculture both because of the growing need to give this sector a profitability comparable with that of industrial and commercial sectors and following the changes which have occurred in eating habits.

In fact just the changes in eating habits and in particular the increased consumption of meat, as well as the search for higher profitability, have stimulated a search for crops which can sustain specialised animal rearing. There has thus been a rapid development in those crops (maize and soya) which are intended in particular for rearing animals, with a gradual abandonment of practices of rotation and association of crops.

In the Po basin, which extends over about 7 million hectares there is the largest nucleus of cultivated soil (3.4 million hectares) in Italy.

10.3 General Picture of the Critical Areas from the Environmental Point of View

In the ambit of the activities aimed at defining actual knowledge about the Po Basin, the MASTERPLAN proposes to give a general picture of the tendency to degradation, identifying those areas considered "critical" from the environmental point of view. Reference is made to a "potential criticality", attributed to those situations which, in theory, make the environment vulnerable to the various types of anthropic transformation of the area. The starting assumption is linked to the obvious relationship of causality which links the anthropic component to the environmental imbalance found in the basin; thus the concept of "criticality" is closely linked to the type and quality of anthropic pressure affecting it.

In the MASTERPLAN complex indicators were constructed which could localise the phenomenon in relation to the existing situations of urban and area development, social and economic changes, and environmental changes and transformations in progress.

The identification of the so-called environmental value is a completely different type of potential criticality. In this case the criticality potential arises from the identification of areas which are in a state of appreciable naturalness, especially if compared with the neighbouring areas of the basin; the state of conservation of these areas should be safeguarded, if possible, and their use should be limited, or their purpose changed.

With regard to future initiatives of clean-up programming and planning, these areas form the reservoir of "naturalness" to be protected against any localised initiatives in or near them which could in some way degrade them.

Three categories of area have thus been identified to represent an aspect of the environmental value:

• protected areas;
• areas subject to hydrogeological constraints and
• wooded areas.

From an ecological point of view, one of the most important areas in the river Po basin is the Delta region.

10.4 The Po Delta Region

From the purely hydraulic point of view, the term Po Delta defines the area directly affected by the 5 active branches of the terminal river system, between the Po di Maistra to the north and the Po di Goro to the south. From the historical point of view, however, this term indicates the whole territory which, from Chioggia to Ravenna, is the result of the dynamic actions of the river, the sea and man over the last 3.000 years: an organic complex of land and water, divided into a vast system of islands, marshes and watercourses, called Polesine since the Middle Ages, which has a completely original appearance [Tomasin, 1990].

A primary characteristic of these lands is their fundamental uniformity: not considering political and administrative divisions, in fact, the whole area from the Adige to the Primaro forms a single territory linked by a series of duney ridges, which over time map out the progressive advances of the coastline. This can be seen from some ancient historians, who identify the Delta as a single lagoon, river and valley environment, organically connected by canals and road networks.

A second element which marks the Polesine is the decisive antinomy between high lands and low lands. Opportunities to cultivate the land and of economic promotion, on the one hand, and hydraulic crises or danger of scarcity on the other, depend critically on differences in height even of just a few centimetres. It is interesting to note that the difference between dry land and irrigated land, typical of regions of Mediterranean Europe, has no meaning in the Polesine, whose historical experience finds comparison only in very small areas of Europe.

A further characteristic of these lands is their extreme dynamism: the whole Polesine complex has always been subject to very strong hydraulic and territorial dynamics, which have a direct and immediate effect on the local populations.

From an economic point of view, the Polesine even now is a peripheral area with respect to the main economic policies of development. It has only been marginally affected by the processes of economic growth which have affected the central and foothill areas. This situation is characterised structurally by [Tomasin, 1990]:

- insufficient growth, with continuous reduction in the number of residents; the loss of the most active and enterprising human forces during the migratory drain of the recent past has heavily penalised the population of the Polesine, where there is a worrying ageing rate and a very insufficient level of training;
- the persistence of a "culture of subjection", imbued with attitudes of passivity and fatalism, which has impeded the diffusion of a different "culture of enterprise" thus hindering the development of entrepreneurial capacities of the local productive class;
- a mainly agricultural economy, dominated from the beginning by large landlords, technically well equipped and extremely efficient;
- an industrial fabric made up of small and medium sized businesses, mainly of craftsmen;
- the presence of considerable environmental resources, historically used for hunting and fishing, which up to now have not been adequately exploited, although there are significant signs of change, mainly the decision of the institution of an inter-regional natural park in the Po Delta.

10.5 The Po Delta Natural Park

The whole area of the Po Delta natural park belongs to two different regions, Veneto and Emilia Romagna, the total surface of the park is 1,456 km^2 (park + buffer zone), 738 km^2 in Veneto and 718 km^2 in Emilia Romagna.

From an environmental point of view, the area of the park presents different interesting aspects. The main characteristics are:

- areas of a great ornithological interest, some of them, according to the *Ramsar Convention*, can be considered of an "international importance";
- different types of flora present in a great variety of ecologically and morphologically distinct natural environments;
- lagoons, wetlands and woods of a great natural interest.

Traditionally, natural parks have four main aims [MacNelly & Miller, 1984]:

(1)   conservation of natural ecosystems,
(2)   recreation,
(3)   scientific research,
(4)   education.

In recent years, to the traditional aims of parks has been added that of the economic and social development of the community involved, which certainly forms the newest and most controversial theme of conservation policy.

Various factors may justify the inclusion of this theme in international debate: the emergence of the economic and financial needs of Third World Countries, where the parks inevitably have connotations of instruments of development; the need of more industrialised countries to reduce their internal imbalances and the connected social tensions, especially in zones of greater natural value; the size of the economic fluxes activated by the parks and, more generally, the political decision to go beyond the dominant utilitarian view of parks.

So far, the need for an improvement of the economic and social conditions of local populations has been satisfied by the principle of compensation, a more modern expression of the old concept of indemnity, which implicitly reveals a limited interpretation of the park; an indemnity is due where, as an activity cannot be carried out, there is a damage [Tomasin, 1990].

From various sides however a new policy is proposed, which promotes development on the same level as conservation and sees the institution of a park as an operation which can have economic ambitions, aimed at a result of financial self-sufficiency, a generator of work and an induced economy. The park, intended as a social investment which can guarantee work and income, would form the factual proof of the compatibility between man and nature, between needs of economic development and environmental conservation.

As an example of such a possibility, we will illustrate an environmental management problem present in a wood, called "boscone della Mesola", part of the area of the Po Delta natural park, where a sharp conflict between environmental and economic aspects seems apparently to exist.

10.6 The "Boscone della Mesola" Environmental Management Problem

Detailed study of the physiographical (paleography, hydrographic network, etc.), physical (geology, pedology, hydrogeology, climatology, hydrology) and biological (vegetation and fauna) characteristics of the Mesola wood has revealed the following facts [IDROSER, 1978]:

(a) the close interdependence between the vegetation and the surface and underground hydrological system,

(b) the considerable natural and ecological importance of a qualitatively and quantitatively well conserved forest ecosystem in an agricultural context strongly affected by man,

(c) the scientific, educational and cultural importance of a considerable vegetational complex, now the last remaining example in the lower Po valley.

The Mesola wood is thus of exceptional value and considerable environmental potential. A range of activities could thus be set up around it which, consistent with the environmental characteristics, could also provide considerable economic and social advantages.

The present Mesola wood formed part of a much greater forest area, which covered a surface of some thousand hectares: already at the end of the 1500s the wood, which up to then was used as park and hunting reserve, was partly cut down and encroached upon. Moreover the pressure exerted by a quickly growing resident population by the intensive exploitation of wood products, the planting of crops in some internal areas and the use as pasture, caused the wood to degrade rapidly. In 1850 the forest area was reduced to about 2200 hectares, while at the beginning of the 1900s it was only 1500 hectares. In following years the forest suffered further because of wars, so that after the second world war the wooded area was reduced to 900 hectares.

In 1954 the State Organisation for State Forests took over ownership of the wood, which was then subjected to operations to restore and reforest areas previously tampered with. The wood's natural equilibrium was however destined to be subjected to a series of severe negative influences. The Po Delta Organisation, formed in 1951, began to set up a project to transform the Delta, which was intended to affect the valleys around the Boscone also. This project to clean up the lagoon valleys completely, was aimed at the setting up of a defence system against sea storms.

Some valleys to the north-west and north-east of the wood, like the Vallona, Totanara and Pioppa valleys, were drained in these years, with negative micro-climatic and water table effects: dry conditions and the drying out of some plants were found in the part of the wood neighbouring the ex-Totanara valley. The project to reclaim the valleys to the west of the wood (Giralda, Gaffaro and Falce) was approved in 1958 by the Ministry of Agriculture and Forests and the work to drain the Giralda and Gaffaro started immediately. The Falce Valley was partially isolated from the Sacco di Goro by the construction of a bank.

In 1969, despite the landscape constraint to which the Falce Valley was subjected from 1963, the Ministry of Agriculture and Forests gave the go-ahead for its exploitation, although prescribing actions to safeguard the wood by irrigation. The drainage works were completed in 1970, but the forest defence works were not started.

The next year the wood began to show some signs of stress in the areas near to the drained valley: there were some drying-out phenomena, which initially killed a few hundred plants (50% of the trees in a 3 hectare area in the Duchesse Park) and damaged very many others, which died in later years. The Ministry of Agriculture then decided to finance measures to restore the equilibrium of the water table, which consisted in the opening of a drainage channel along the Falce Valley, cleaning up the drainage network in the wood, the supply of water to the Boscone by the reactivation of the Balanzetta pumping station and the excavation of the channel to supply the "Canal Bianco". The mortality of plants became less although the wood continued - and continues even today - to show signs of stress (it has been evaluated that 120.73 hectares of wood have been damaged). Because of this, a project of flooding the Falce Valley has been proposed by the technicians of IDROSER [1985]. It should be noted that if the flooding project will be carried out, it is necessary to find out new economic activities able of substituting the agricultural land use. The most promising one is fish rearing.

10.7 Flooding of the Valley Area by Building Independent Basins with Different Salt Level

Among clay valleys, the most biologically active and productive are those with efficient water exchange guaranteed by a rational distribution of supply and discharge outlets, allowing their salt level to be regulated as necessary. The project to use the Falce Valley for fishrearing must thus be oriented towards this type of solution.

The project of using initially about half of the surface of the Falce valley area, which would be divided into four basins, three of which would be filled with fresh water (about 20 hectares each) and one with brackish water (About 50 hectares), could represent either a permanent solution or a partial and transient solution, set up with the aim of obtaining clear information about the productive possibilities of the valley area and the response of the surrounding environment to gradual reflooding of the whole valley.

In order to implement this project, the first action to be undertaken is the construction of a longitudinal bank parallel to the Falce channel, then other three banks perpendicular to that should be constructed thus creating the four basins described above.

From the fish rearing point of view the most important role in the three basins with fresh water must be given to carp (Cyprinus carpio L.), together with other species, of which the herbivorous carp (Ctenopharyngodon idellus Val.), which has recently been introduced into Italy, would seem particularly suitable. Together with carp it improves production.

The association of the two species would allow the carp, after one or two years, subject to rational initial seeding, to become self-sufficient and in later years to supply considerable production of adult and young fish. As the herbivorous carp does not reproduce in Italian waters it must be introduced annually or every two years and would play the important role of containing water infestations, as well as supplying a product commercially very similar to carp. Together with the two species mentioned one could also introduce two species of grey mullet (Mugil cephalus and Liza ramada) into the three basins. It can live and grow regularly in fresh water.

Eel could also form part of the polycultural system. The rearing of this species in fresh water, although it could be caught almost exclusively using fixed or mobile nets, seems more justified than in brackish water because there is no risk of attack by the parasite Argulus giordanii which cannot live in fresh water.

In the basin next to the "Taglio della Falce" weir, supplied with brackish water from the Sacca di Goro, one could rear some of the fish species: the most suitable would seem to be grey mullet, sea bass and gilthead while eel would seem to be less suitable because of the difficulties in catching it.

From the hydraulic point of view this solution would be without problems, as the single basins are small there would not be particular water circulation difficulties, all the more so as the entire amount of fresh water available could be poured into each basin in turn.

From the economic point of view, indications of the presumed management costs and the yields for the assumed fish rearing uses are very approximate; here the uncertainties on use of labour, productivity of the valley, price of the various species of fish (from the point of view both of buying young fish and selling the product) are so large that the figures given should be taken conditionally. Important factors to be borne in mind are the incidence of epidemic diseases or water pollution, which can reduce the quantity and quality of the catch yield drastically.

There is no doubt that the flooding of the Falce valley will have excellent environmental consequences, in particular, on the ecological equilibrium of the wood. It has been evaluated that the flooding of Falce valley will modify the present negative trend of the wood starting from the second year, and the natural growth rate will be fully reached starting from the 9-th year. However, one should note that a conflict between ecological and economic ends, seems to exist: the flooding of the valley implies the lost of actual agricultural production with serious bad repercussions in terms of income and employment on the whole area. As a consequence, it is necessary to explore the possibility of finding a kind of economic-ecological equilibrium, i.e. economic activities substituting the agricultural one, compatible with the flooding of the valley. Thus, a comparison between the future possible economic scenario and the so called "option-zero" (business as usual) has to be considered. This will be carried out in the next Chapter, where the possibility of an ideal optimal agriculture activity will also be considered.

Our study is mainly based on information provided by IDROSER. However, originally only traditional monetary evaluation methods have been used, thus we will re-structure the whole problem in function of a qualitative multicriteria evaluation.

CHAPTER 11

THE PO DELTA NATURAL PARK: THE "BOSCONE DELLA MESOLA" ENVIRONMENTAL MANAGEMENT PROBLEM

11.1 The Productive Use of the Exploited Land of the Falce Valley

11.1.1 The Economic Result of the Cultivation of the Falce Valley with the Present Agricultural Practice Criteria

On the basis of the data supplied by the present Falce Valley farmers (checked with privileged observers from the Ferrara Agricultural Inspectorate and INEA) the agricultural operation balance has been reconstructed by IDROSER.

Concerning the crops, productivity and gross yields the most optimistic indications were accepted, neglecting the doubts on crop yields expressed by some observers. The productive structure considered has a particularly low level of machinery, to the point that it should be considered that all the mechanical operations are carried out by third parties. The depreciation concerns current operating equipment and an average incidence of 150,000 lire per hectare was assumed[1] .

The labour necessary is limited to vegetable and fruit growing (as mechanised techniques are not adopted) and amounts to 35% of the gross marketable production. Given a total labour cost of 140.333 million lire 8 workers are employed on average.

The financial costs have been evaluated at 5% of the gross marketable production. The ratio between added value and total of yields is assumed at 58% and is taken from the indications of the experts with reference to the characteristics of the area concerned. Subtracting the total costs from the added value one finds that the gross result of management produced from agricultural use of the Falce Valley in current conditions is about 64 million lire per year (the total yields being L. 797,850,000).

This yield represents the profit of the whole operation following the exploitation in the present state of the area, and represents the social cost to be

---

[1] All prices are considered in 1985 Italian lire; if it is not differently specified, they refer to the whole Falce Valley area.

imputed to the balance when evaluating the public profitability of the flooding project in relation to the loss of cultivated land.

11.1.2 The Improvement of Crop-Growing Methods and of Results of Agricultural Management

An assumption of optimised agricultural practice to quantify the maximum social profit which could be made from the present productive destination can be formulated. The crop-growing system thus predicted is characterised by:

(1) use of an irrigation system for the whole cultivated area which results in a greater burden of depreciation (estimated at 300,000 lire per ha);

(2) four-yearly crop-rotation cycle based on grain, maize, beetroot and vegetables and fruit in equal measure with respect to the area cultivated (61.25 ha for each cultivation);

(3) introduction of regeneration crops - forage - in the summer for half of the area cultivated with grain in the winter;

(4) introduction of a doubling of the use of a third of the area intended for growing vegetables and fruit based on an integration of summer and winter crops;

(5) increase of unit yields following irrigation;

(6) maintenance of the use of third parties for mechanical working and use of the direct work force for vegetable and fruit growing for which it is calculated 20 workers are employed; evaluation of the average incidence of 5% of the financial burden and 58% of the added value on the total of yields.

Assuming the criteria stated above the gross profit which can be obtained, in the best conditions, by cultivation of the land of the Falce Valley is about 159 million lire per year (the total yields being L. 1,462,497,000).

11.2 New Economic Scenario

The evaluation of the gross result of management of fish rearing is derived from:

(1) calculation of the gross yields based on the quantity of fish produced;

(2) determination of the added value as 58% of the gross yields as indicated in the national averages of agricultural and fish-rearing activities;

(3) subtraction of depreciation (indicated at 150.000 lire per ha) connected with agricultural machinery, labour (indicated at 25% of the yields and consistent

at 6 workers in normal conditions) and financial burdens (estimated at 5% of the yields).

In normal operating conditions fish rearing in the Falce Valley can produce a gross management result of about 136 million lire per year (considering a progressive flooding of the whole area).

As part of a project covering a wider area, other social and economic aspects should be considered, particularly:

(1)   the effect on the area of investments for the work needed for the flooding project and the associated multiplicative effects of the expense which, as will be seen below, translate into an increase of work;

(2)   evaluation of the advantages of tourists, either specifically to the Boscone (visitors) or because of the flooding (sport fishing, recreation area), would be entrusted to a plan of rationalisation and integration of the possibilities and potential of the Delta Park area; in this context one could quantify the economic and work benefits connected on the one hand with improvement and expansion of tourist attractions and on the other with the effects of the de-congestion of the natural reserve with respect to the excessive present pressure of visitors;

(3)   as part of the activities described above, conditions would be created for expansion of the jobs available, allowing the reabsorbtion of people who would be out of work following the modification of economic activities (from agriculture to fish rearing).

In particular the modification of the jobs available following the Falce Valley flooding project can be summarised in a first approximation in the following terms.

- The work of digging and building banks, etcetera, which would be necessary for the project, would bring money and work to the areas around the Boscone. Assuming as an example that 20% of the engineering work can be translated as a benefit for the area, one finds that in 3 years of engineering work, work is produced for an average of 9 people per year.
- Cultivation of the exploited lands is very backward, while the prospects opened by fish rearing would allow better exploitation of the area supplying higher yields per hectare than farming.

Moreover the rearing of fallow deer is being studied, to produce and sell high-quality meat, which would reduce the present state of overpopulation of these animals in the Boscone della Mesola.

The work force employed in present farming of 240 hectares of the Falce Valley is evaluated on the annual basis of 8 people and should be compared with the stable work force which would be needed for the fish rearing, which would be 6 people. Initially about 300 animals must be removed from the Boscone and form the animal heritage on which rearing would be based (tending to reproduce at a rate of 1.7% per year), which would require 3 new workers. One job is thus created.

11.3 Evaluation of Different Possible Courses of Action

11.3.1 Efficiency: Economic Analysis of Different Activities

This analysis will be carried out from a rigourously economic point of view. Three different alternative possibilities are evaluated:
(1)   business as usual,
(2)   optimised agricultural practice,
(3)   fish rearing.

Two evaluation criteria are used:
(1)   gross profit,
(2)   employment.

A) GROSS PROFIT:
(1)   business as usual: about 64,000,000 Lire per year,
(2)   optimised agricultural practice: about 159,000,000 Lire per year,
(3)   fish rearing: because of the progressive flooding of the valley and the characteristics of the fish, the full production will be reached only in the 7-th year of production. The annual gross profits can be estimated as follows:
1-th year: 6,176,000
2-th year: 18, 962,000
3-th year: 29,466,000
4-th year: 36,984,000
5-th year: 42,630,000
6-th year: 45,099,000
7-th year: 135,646,000

As one can see, from a financial point of view, the optimised agriculture practice results to be the best. However, between current agriculture practice and fish rearing, if a medium time horizon is taken into account, fish rearing results to be the better option.

B) EMPLOYMENT:

(1)   business as usual: 8 men/year,

(2)   optimised agricultural practice: 20 men/year,

(3)   fish rearing: 6 men/year.

It is clear that also in this case the optimised agriculture practice results the best alternative. Fish rearing results the worst alternative; however, it should be noted that the difference with the current agriculture practice is not so big.

As a consequence of this brief analysis, two main considerations can be done:

• if fish rearing substitutes agricultural activity, the project of flooding the Falce valley can be acceptable also from a strictly economic point of view. However, the optimising of the agriculture practice seems to ensure a better overall performance of the economy;

• the above economic analysis is essentially a partial analysis. This holds true because of two main reasons:

    (a) all the economic consequences deriving from the integration of the area into the Po Delta natural Park have not been considered,

    (b) the environmental issues have been completely neglected, no trade-off between ecological and economic aspects has been considered.

In the next section a comprehensive economic-environmental evaluation of alternative courses of action will be carried out by means of multicriteria analysis.

11.3.2   Economy-Environment Interactions: a Fuzzy Multicriteria Evaluation of Five Possible Courses of Action

The alternatives taken into consideration are:

(1)   business as usual,

(2)   optimised agriculture,

(3)   flooding of the Falce Valley,

(4)   partial flooding in combination with business as usual,

(5)   partial flooding in combination with optimised agriculture.

The following set of evaluation criteria can be used:

(1)   Gross profit

There are no problems for the values connected with business as usual and optimised agriculture practice, on the contrary, if the flooding project is evaluated from a broader point of view, i.e. the multiplicative effects of investments for the flooding project, the integration with the whole Po Delta natural park and the fallow deer rearing are also considered together with the fish rearing the evaluation becomes much more difficult. The following assumptions are made:

•   the returns coming from the integration with the natural park are not considered here; a separate criterion called "tourist attractiveness" is used without any monetary quantification;

•   the fallow deer rearing seems a quite interesting complementary economic activity to fish rearing. It has been evaluated [IDROSER, 1985] that it is possible to keep about 10 - 12 animals per hectare and that an initial animal heritage of 300 animals (with a reproduction rate of 1.7%) can be assumed. If also the multiplicative effects of investments are taken into account, it is reasonably to assume that the value considered before, of about 136,000,000 Lire per year could increase by a certain percentage. By making the non too optimistic hypothesis of a maximum increase of 10%, a maximum gross profit of about 150,000,000 per year is obtained. This range can be represented by a fuzzy number like the one of Figure 11.1; the Gaussian membership function of this fuzzy number is the following:

$$e^{-0.077\,(x\,-\,143)^2} \tag{11.1}$$

For the cases in which the hypothesis of partial flooding is considered, the assumption that roughly a little bit less than the 50% of the values is a good approximation of reality is made. In Figures 11.2 and 11.3 the fuzzy numbers "approximately 95,000,000" and "approximately 147,000,000" are illustrated; as one can see in defining the shape of the membership function a cautionary approach has been adopted. Analitically, these fuzzy numbers have the following membership functions (arbitrarily defined on the basis of the above heuristic considerations):

$$\begin{cases} 1 - e^{-408\left(\frac{x-85}{85}\right)^2} & \text{if } 85 \leq x < 95 \\ 1 & \text{if } x = 95 \\ \left[e^{-150\left(\frac{x-95}{95}\right)^2}\right]^2 & \text{if } 95 < x < +\infty \end{cases} \qquad (11.2)$$

$$\begin{cases} 1 - e^{-1500\left(\frac{x-140}{140}\right)^2} & \text{if } 140 \leq x < 147 \\ 1 & \text{if } x = 147 \\ \left[e^{-470\left(\frac{x-147}{147}\right)^2}\right]^2 & \text{if } 147 < x < +\infty \end{cases} \qquad (11.3)$$

It has to be noted that given the deep uncertainties in these evaluations, the use of fuzzy sets seems very appropriate.

(2)   Employment

For the flooding project the summation of fish rearing and fallow deer rearing is taken into account. The employment deriving from the engineering works is not taken into account.

(3)   Tourist attractiveness

(4)   Recreational attractiveness

This criterion is connected to the function of environment of supplying utility directly in the form of aesthetic and spiritual comfort; this is the reason why it is kept separate from tourist attractiveness, that here is considered only an economic point of view.

(5)   Ecological equilibrium of the wood

(6)   Risk of causing ecological damages.

Since the use of fertilisers is a common practice in agriculture, this economic activity normally presents more environmental risks than aqua culture. It has to be noted that in the case of partial flooding, the possibility of bad environmental consequences of agriculture increases because both the wood and the aquatic ecosystems can be affected.

It should be noted that the investment costs are not considered in the list of evaluation criteria. This because it is assumed that society is willing to spend the money needed to establish the ecological equilibrium of the wood (in other

220

words the monetary value of the wood is at least equal to the investment cost). Thus the problem is to find out if the economy of the area is depressed by the project (thus a compensation is needed), or if in a medium period, a kind of economic-ecological equilibrium can be reached (sustainable in the long run), thus showing that environmental policy can have positive consequences also on the economc side. In any case, one has to note that within a business as usual framework, 50,000,000 Lire per year is spent for irrigation of the wood. Moreover, in a monetary evaluation framework, the investment cost should be compared with the monetary damage to the wood.

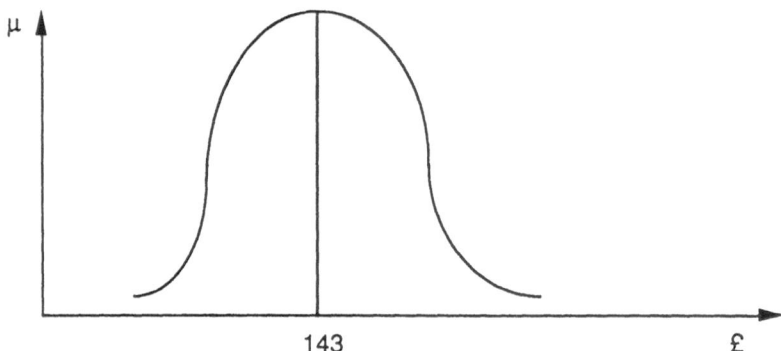

Figure 11.1   Symmetric Representation of the Fuzzy Number "Approximately 143,000,000"

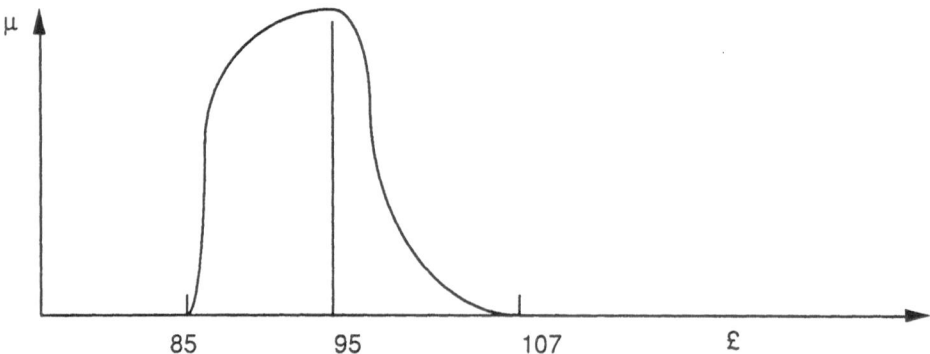

Figure 11.2   Representation of the Fuzzy Number "Approximately 95,000,000"

The impact matrix related to the "boscone della Mesola" problem is shown in Table 11.1.

As one can see, this evaluation matrix is mixed in nature. In particular, on criterion one both crisp numbers and fuzzy numbers are present. As an example of comparison between two overlapping fuzzy numbers, in Appendix 1 the fuzzy relations obtained by comparing "approximately 143,000,000" and "approximately 147,000,000" are shown.

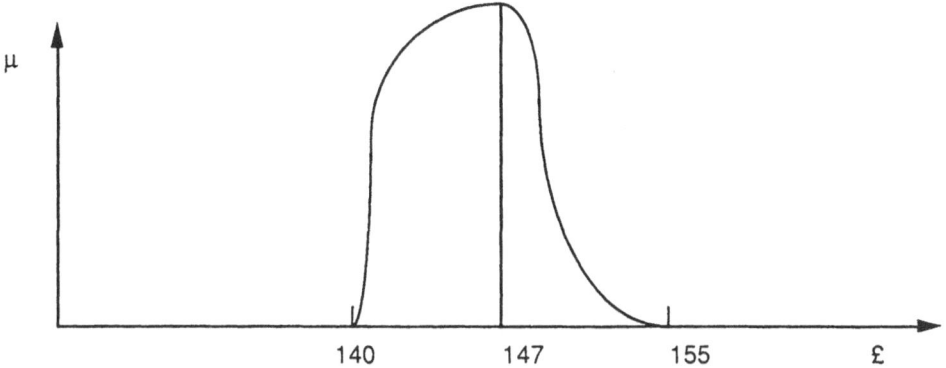

Figure 11.3   Representation of the Fuzzy Number "Approximately 147,000,000"

criteria	units	business as usual (a)	optimised agriculture (b)	flooding (c)	partial flooding + current agriculture (d)	partial flooding + optimised agriculture (e)
g_1	Italian Lire	64,000,000	159,000,000	approximately 143,000,000	approximately 95,000,000	approximately 147,000,000
g_2	men/year	8	20	9	8	14
g_3	linguistic	bad	bad	good	moderate	moderate
g_4	linguistic	moderate	moderate	good	moderate	moderate
g_5	linguistic	bad	bad	good	good	good
g_6	linguistic	moderate	bad	good	bad	bad

Table 11.1   A Fuzzy Impact Matrix for Different Land Uses of the Falce Valley

Given the nature of the information contained in this impact matrix, the use of the NAIADE method seems very appropriate. By applying NAIADE the degrees of truth of a linguistic evaluation described in Appendix 2 are obtained; the min operator is used.

By computing the ϕ^+ and ϕ^- indices by using the min operator, the following results are obtained:

α=0.20

$\phi^+(a)=0.124$ $\phi^+(b)=0.715$ $\phi^+(c)=1$ $\phi^+(d)=0.357$
$\phi^-(a)=0.899$ $\phi^-(b)=0.609$ $\phi^-(c)=0.336$ $\phi^-(d)=0.633$

$\phi^+(e)=0.784$
$\phi^-(e)=0.401$

Then according to ϕ^+ the following ranking is obtained
c→   e→   b→   d→   a
and according to ϕ^- the following ranking is obtained
c→   e→   b→   d→   a
Therefore, the intersection gives the same result

α=0.40

$\phi^+(a)=0.073$ $\phi^+(b)=0.442$ $\phi^+(c)=0.758$ $\phi^+(d)=0.225$
$\phi^-(a)=0.688$ $\phi^-(b)=0.344$ $\phi^-(c)=0.197$ $\phi^-(d)=0.395$

$\phi^+(e)=0.494$
$\phi^-(e)=0.201$

Therefore, the same ranking is again obtained.

α=0.60

$\phi^+(a)=0.011$ $\phi^+(b)=0.256$ $\phi^+(c)=0.390$ $\phi^+(d)=0.095$
$\phi^-(a)=0.414$ $\phi^-(b)=0.174$ $\phi^-(c)=0.096$ $\phi^-(d)=0.195$

$\phi^+(e)=0.241$
$\phi^-(e)=0.074$

Then according to ϕ^+ the following ranking is obtained
c→   b→   e→   d→   a
and according to ϕ^- the following ranking is obtained

e→   c→   b→   d→   a

Therefore, the intersection gives the following final result:

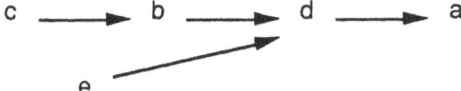

For both α= 0.20 and α= 0.40, the same complete preorders are obtained. For α=0.60, a partial preorder is obtained; the ranking is equal to the previous ones with the exclusion of action **e** that now results incomparable with **c** and **b**, no rank reversal is present. By looking to the pairwise linguistic evaluations, it is possible to see that for α=0.40, it is almost never possible to discriminate among the alternatives. For α= 0.20, **c** results better than **e**, but also the credibility of the preference of **e** over **c** seems high; between **b** and **e**, the relation of indifference has the highest credibility, both the preference of **b** over **e** and vice versa are high (the preference of **b** over **e** is even higher, then the preference of **e** over **b** in the final ranking is mostly due to the overall relationship among the alternatives). For α=0.40, it is not possible to discriminate both between **b** and **e** and between **c** and **e**, therefore we can conclude that the relationship of incomparability obtained for α=0.60 between **e** and the pair **c** and **b** is very plausible. On the contrary, the evaluation of the pair **b** and **c** is very clear, for both α= 0.20 and α=0.40, the preference of **c** over **b** is quite clear. Finally, it is evident that **a** and **d** are the worst actions.

In conclusion, from an economic-environmental perspective it is not possible to defend the business as usual option (**a**), also the mix between current agriculture and flooding (**d**) is very weak. The flooding option (**c**) seems to be better than the ideal optimised agriculture situation (**b**). The mix between optimised agriculture and flooding (**e**) is difficult to compare with both optimised agriculture and flooding; however, since flooding looks better than optimised agriculture, the final decision should be made between flooding and the mix between optimised agriculture and flooding.

It has to be noted that this analysis does not take into account equity considerations. This will be done in the next section, where, although in a rough way, also intergenerational equity will be considered.

11.3.3   Equity: an Intra-Inter Temporal Conflict Analysis

Following the philosophy of the planning balance sheet method, the matrix presented in Table 11.2 shows the impacts of the different courses of action on different income/interest groups. It appears that the information concerning the diverse plan impacts is rather inaccurate; the degree of uncertainty on the impacts of the plans is high, so that a representation of such impacts in fuzzy terms seems very appropriate. The impacts have been evaluated on the basis of heuristic principles, by interacting with people of the region and some experts.

Six main interest groups can be distinguished:
- Farmers
- Environmentalists
- Recreationers
- Landless Labourers
- Residents in the Po Delta Area
- Future Generations

In order to elaborate this information with a normative purpose, the fuzzy clustering algorithm presented in Section 8.5 will now be used.

Interest Groups	Alternatives				
	business as usual (a)	optimised agriculture (b)	flooding (c)	partial flooding + current agriculture (d)	partial flooding + optimised agriculture (e)
Farmers (1)	good	very good	very bad	bad	moderate
Environmentalists (2)	bad	bad	very good	moderate	moderate
Recreationers (3)	bad	bad	good	good	good
Landless Labourers (4)	moderate	moderate	good	good	good
Residents in the Po Delta Area (5)	bad	bad	good	moderate	moderate
Future Generations (6)	bad	moderate	very good	moderate	good

Table 11.2   Fuzzy Evaluations of Impacts of Different Alternative Courses of Action on Different Interest Groups

By applying the semantic distance described in (6.70) with p=2, after the transformation s= 1/1+d, the following similarity matrix for all possible pairs of interest groups is obtained:

	1	2	3	4	5	6
1	1	0.438	0.440	0.478	0.465	0.448
2	0.438	1	0.468	0.579	0.8	0.675
3	0.440	0.468	1	0.675	0.675	0.648
4	0.478	0.579	0.675	1	0.595	0.648
5	0.465	0.8	0.675	0.595	1	0.648
6	0.448	0.675	0.648	0.648	0.648	1

By using the notion of max-min composition, the following new fuzzy relations are derived:

$$R^2$$

	1	2	3	4	5	6
1	1	0.478	0.478	0.478	0.478	0.478
2	0.478	1	0.675	0.648	0.8	0.675
3	0.478	0.675	1	0.675	0.675	0.648
4	0.478	0.648	0.675	1	0.675	0.648
5	0.478	0.8	0.675	0.675	1	0.675
6	0.478	0.675	0.648	0.648	0.675	1

$$R^3$$

	1	2	3	4	5	6
1	1	0.478	0.478	0.478	0.478	0.478
2	0.478	1	0.675	0.675	0.8	0.675
3	0.478	0.675	1	0.675	0.675	0.675
4	0.478	0.675	0.675	1	0.675	0.675
5	0.478	0.8	0.675	0.675	1	0.675
6	0.478	0.675	0.675	0.675	0.675	1

Since in the series of max-min compositions $R^3=R^4$, the transitive closure is $\mathcal{R}= R \cup R^2 \cup R^3 = R^3$

226

Since $\mathcal{R}$ is a similitude relation, it can be decomposed into equivalence classes with respect to the degree of similarity α.

Thus the application of the clustering procedure leads to the following results (see Figure 11.4). The results of the coalition formation process are very clear. The interests of environmentalists (2) and of the residents in the Po Delta area (5) bear a very close correspondence. Recreationers (3), landless labourers (4) and the special "interest group" of future generations (6) join quite soon the coalition, thus no serious conflict seems to exist and therefore the probability of finding a compromise solution is high. The interest group of farmers (1) presents a more individualistic character, unreconciliable differences between this group and all the others seem to exist.

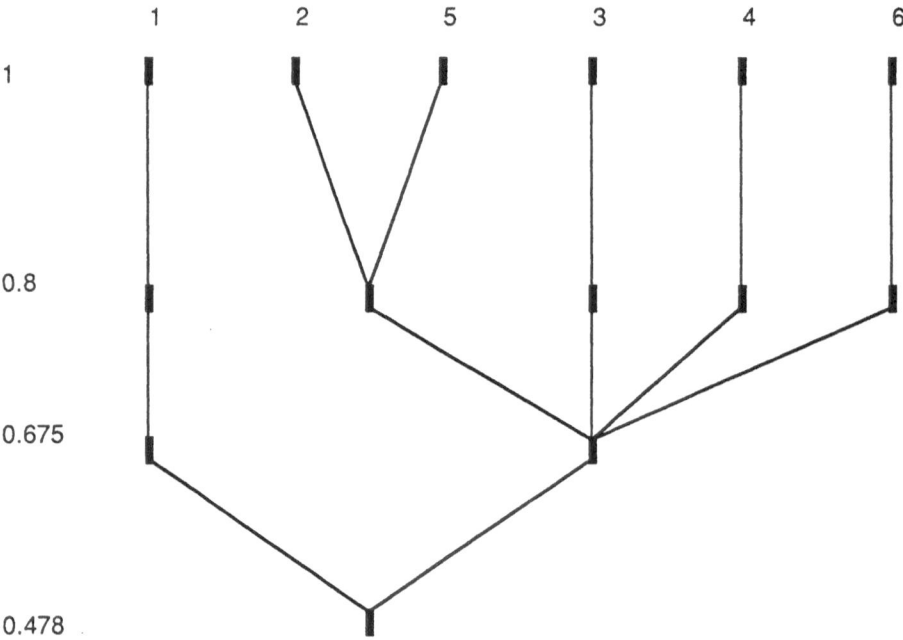

Figure 11.4   Dendrogram of the Coalition Formation Process

The alternative that presents the best impact on environmentalists and residents in the Po Delta area is the flooding project. On recreationers, landless labourers and future generations all the different types of flooding projects (complete or partial) have good impacts, however overall the complete flooding seems slightly preferable to the other options. Moreover, it has to be noted that the mix between flooding and current agriculture is not defendable from an economic-environmental point of view. Farmers are the only ones taking advantages from "business as usual" or optimised agriculture options, unfortunately both courses of action are not defendable.

Finally, it is possible to conclude that only the flooding or the partial flooding plus optimised agriculture options are completely defendable (from economic, environmental and equity points of view). The complete flooding project seems slightly better (above all from an environmental point of view), however the mix between optimised agriculture and flooding is the only project that minimises the conflict with the interest group of farmers.

11.4 Conclusions

From the experience made in this case study, it is possible to draw some conclusions. First of all, one has to admit that the use of fuzzy sets can be a very useful tool in modelling environmental management problems characterised by deep uncertainties and approximate evaluations. As a consequence, the development of the NAIADE method seems a meaningful step for improving environmental management techniques. According to MCDA philosophy, no optimal solution has been found, while some actions considered defendable have been identified.

A major point is that these actions can in principle be identified by taking into account all three conflictual values of economics, viz. efficiency, equity and sustainability. In particular, given the complexity inherent in the concept of sustainability, this has been operationalized in a rough way. Just by taking into considerations both economical and environmental consequences of the problem; intergenerational equity has been considered too. Of course, this is a second best approach, but given the state of the art, our approach seems to be a meaningful undertaking in an operational framework.

Finally, one should note that the institution of a natural park may have positive consequences also from the economic point of view, if economic activities more compatible with the natural environment are carried out.

APPENDIX 1

An Example of Fuzzy Relations Obtained for Two
Overlapping Fuzzy Numbers

Let us take into consideration the following two fuzzy numbers:

$$A = e^{-0.077\,(x-143)^2}$$

$$B = \begin{cases} 1 - e^{-1500\left(\frac{x-140}{140}\right)^2} & \text{if } 140 \leq x < 147 \\ 1 & \text{if } x = 147 \\ \left[e^{-470\left(\frac{x-147}{147}\right)^2}\right]^2 & \text{if } 147 < x < +\infty \end{cases}$$

The following fuzzy relations are obtained;

$$\begin{cases} \mu_{>>} = 0 \\ \mu_{>} = 0 \\ \mu_{\cong} = 0.41 \\ \mu_{=} = 0.10 \\ \mu_{<} = 0.79 \\ \mu_{<<} = 0.99 \end{cases}$$

The following values for the parameters c have been used:

$c_{>}$ 5

$c_{>>}$ 0.011

$c_{\cong}$ 0.138

$c_{=}$ 0.056

While for the fuzzy relations $\mu_{<}$ and $\mu_{<<}$ the difference between the expected values of B and A have been used (E(B) - E(A)$\cong$ 4.368); for the fuzzy relations $\mu_{\cong}$ and $\mu_{=}$ the semantic distance in (6.20) has been used. Then it is

$$S_d (B, A)= \int_{147}^{+\infty} \int_{-\infty}^{+\infty} |x - y| \frac{\left[e^{-470 \left(\frac{x - 147}{147}\right)^2}\right]^2}{3.816} \circ \frac{e^{-0.077 (x - 143)^2}}{6.349} \, dydx +$$

$$+ \int_{140}^{147} \int_{-\infty}^{+\infty} |x - y| \frac{1 - e^{-1500 \left(\frac{x - 140}{140}\right)^2}}{4.249} \circ \frac{e^{-0.077 (x - 143)^2}}{6.349} \, dydx \cong 6.329$$

APPENDIX 2

Pairwise Linguistic Evaluations

a is better than b	$\tau=0$	
a and b are indifferent	$\tau=1$	$(\alpha=0.20)$
a is worse than b	$\tau=0.694$	
a is better than b	$\tau=0$	
a and b are indifferent	$\tau=0.464$	$(\alpha=0.40)$
a is worse than b	$\tau=0$	
a is better than b	$\tau=0$	
a and b are indifferent	$\tau=0$	$(\alpha=0.60)$
a is worse than b	$\tau=0$	
a is better than c	$\tau=0$	
a and c are indifferent	$\tau=0.261$	$(\alpha=0.20)$
a is worse than c	$\tau=1$	
a is better than c	$\tau=0$	
a and c are indifferent	$\tau=0$	$(\alpha=0.40)$
a is worse than c	$\tau=1$	
a is better than c	$\tau=0$	
a and c are indifferent	$\tau=0$	$(\alpha=0.60)$
a is worse than c	$\tau=0$	

a is better than d	$\tau=0$	
a and d are indifferent	$\tau=1$	$(\alpha=0.20)$
a is worse than d	$\tau=1$	
a is better than d	$\tau=0$	
a and d are indifferent	$\tau=0$	$(\alpha=0.40)$
a is worse than d	$\tau=0.271$	
a is better than d	$\tau=0$	
a and d are indifferent	$\tau=0$	$(\alpha=0.60)$
a is worse than d	$\tau=0$	
a is better than e	$\tau=0$	
a and e are indifferent	$\tau=0.451$	$(\alpha=0.20)$
a is worse than e	$\tau=1$	
a is better than e	$\tau=0$	
a and e are indifferent	$\tau=0$	$(\alpha=0.40)$
a is worse than e	$\tau=1$	
a is better than e	$\tau=0$	
a and e are indifferent	$\tau=0$	$(\alpha=0.60)$
a is worse than e	$\tau=0.411$	
b is better than c	$\tau=0.710$	
b and c are indifferent	$\tau=0$	$(\alpha=0.20)$
b is worse than c	$\tau=1$	
b is better than c	$\tau=0$	
b and c are indifferent	$\tau=0$	$(\alpha=0.40)$
b is worse than c	$\tau=1$	
b is better than c	$\tau=0$	
b and c are indifferent	$\tau=0$	$(\alpha=0.60)$
b is worse than c	$\tau=0.031$	
b is better than d	$\tau=0.677$	
b and d are indifferent	$\tau=0.874$	$(\alpha=0.20)$

b is worse than d $\qquad \tau=0.564$

b is better than d $\qquad \tau=0$
b and d are indifferent $\qquad \tau=0 \qquad (\alpha=0.40)$
b is worse than d $\qquad \tau=0$

b is better than d $\qquad \tau=0$
b and d are indifferent $\qquad \tau=0 \qquad (\alpha=0.60)$
b is worse than d $\qquad \tau=0$

b is better than e $\qquad \tau=0.807$
b and e are indifferent $\qquad \tau=1 \qquad (\alpha=0.20)$
b is worse than e $\qquad \tau=0.66$

b is better than e $\qquad \tau=0$
b and e are indifferent $\qquad \tau=0 \qquad (\alpha=0.40)$
b is worse than e $\qquad \tau=0$

b is better than e $\qquad \tau=0$
b and e are indifferent $\qquad \tau=0 \qquad (\alpha=0.60)$
b is worse than e $\qquad \tau=0$

c is better than d $\qquad \tau=1$
c and d are indifferent $\qquad \tau=0.904 \qquad (\alpha=0.20)$
c is worse than d $\qquad \tau=0$

c is better than d $\qquad \tau=0.837$
c and d are indifferent $\qquad \tau=0.281 \qquad (\alpha=0.40)$
c is worse than d $\qquad \tau=0$

c is better than d $\qquad \tau=0$
c and d are indifferent $\qquad \tau=0 \qquad (\alpha=0.60)$
c is worse than d $\qquad \tau=0$

c is better than e $\qquad \tau=1$
c and e are indifferent $\qquad \tau=0.694 \qquad (\alpha=0.20)$
c is worse than e $\qquad \tau=0.857$

c is better than e	$\tau=0$	
c and e are indifferent	$\tau=0$	$(\alpha=0.40)$
c is worse than e	$\tau=0$	
c is better than e	$\tau=0$	
c and e are indifferent	$\tau=0$	$(\alpha=0.60)$
c is worse than e	$\tau=0$	
d is better than e	$\tau=0$	
d and e are indifferent	$\tau=1$	$(\alpha=0.20)$
d is worse than e	$\tau=0.77$	
d is better than e	$\tau=0$	
d and e are indifferent	$\tau=0.884$	$(\alpha=0.40)$
d is worse than e	$\tau=0$	
d is better than e	$\tau=0$	
d and e are indifferent	$\tau=0.284$	$(\alpha=0.60)$
d is worse than e	$\tau=0$	

CHAPTER 12

CONCLUSIONS: RETROSPECT AND PROSPECT

12.1   Summary

The growth of world population and the rapid growth of economic activity have caused environmental stress in all socio-economic systems. The awareness of actual and potential conflicts between economic progress in production, consumption, and technology and the environment has led to the concept of *"sustainable development"*.

Given the complexity inherent in the concept of sustainable development, any method trying to operationalize this in a planning context, can be considered a kind of "second best". This is the main reason why the less ambitious concept of "environmental management" has been preferred here.

In order to operationalize environmental management in a regional context, issues such as economic-ecological integration, multiple use, inter-regional spatial links and trade-offs, and uncertainty are of a fundamental importance. In Part A, it has been argued that most of these issues can be tackled by multicriteria evaluation in an efficient way. Since multicriteria techniques are based on a "constructive" rationality and allow one to take into account conflictual, multidimensional, incommensurable and uncertain effects of decisions, they can be considered perfectly consistent with the methodological foundations of ecological economics.

Since spatial-environmental systems are complex systems characterised by subjectivity, incompleteness and imprecision, Part B deals with multicriteria evaluation in a fuzzy environment; since the comparison of fuzzy sets is one of the most important open problems, a new approach based on the use of areas instead of traditional intersections has been illustrated. This approach is also suitable for the problem of equivalence of the procedures used in order to standardise different kinds of criterion scores typical of qualitative multicriteria evaluation. Thus, a new qualitative multicriteria method, the so-called NAIADE method has been developed. Its main characteristic is the possibility of dealing with crisp, fuzzy and stochastic criterion scores simultaneously.

A fuzzy conflict resolution procedure aimed to be integrated with the NAIADE method has also been presented; by means of this procedure equity issues may also be taken into consideration.

In Part C, all the developed mathematical procedures have been applied to a real-world environmental management problem.

12.2   Main Results of the Present Study

The present study had 4 main objectives:
(1) to show the overall good theoretical and practical performance of multicriteria methods in tackling environmental problems. In particular, the consistency between multicriteria evaluation and the epistemological foundations of ecological economics is deeply investigated;
(2) to develop a new multicriteria evaluation method able to deal with crisp, stochastic and fuzzy criterion scores, in order to tackle economy-environment interactions;
(3) to develop a conflict analysis procedure aimed to be integrated with multicriteria methods, in order to operationalize equity issues;
(4) to show the empirical behaviour of such mathematical procedures by means of illustrative and real world environmental management problems.

The first objective has been tackled in Part A; the following conclusions may be drawn:
(1)   Natural life-supporting ecosystems are negatively affected by the disposal of wastes from the economic system. If the economy-environment interactions are taken into account, immediately a broad question about the capability of the natural environments to sustain the economy araises.
(2)   Substitution of more productive man-made capital for natural capital is not an acceptable answer to environmental problems.
(3)   The idea of maintaining the natural capital stock is important and desirable; unfortunately it is very difficult to operationalize. Its main problem is connected to the possibility of valuating environmental goods in money terms. If such a valuation were possible, sustainability could be operationalized by taking into account a portfolio of projects. In this case, individual projects may use environmental resources as long as this is compensated elsewhere in the set of projects.
(4)   Environmental problems are very complex and characterised by scientific uncertainty. Any method trying to operationalize the concept of sustainable development is necessarily a second best approach.
(5)   In economic theory three main conflictual values can be identified: allocation, distribution and scale. In an operational framework, this means that an

exhaustive analysis has to take into consideration efficiency criteria, ethical criteria and ecological criteria, thus a multidimensional paradigm is needed.

(6)   Ecological economics explicitly recognises that economy-environment interactions are also characterised by significant institutional, political, cultural and social factors through which action is carried out. The use of several evaluation criteria is desirable. This implies that in the framework of ecological economics, the maximisation and the weighting premises of neo-classical economics have also to be changed.

In Chapters 3 and 4, the main characteristics of cost-benefit analysis and multicriteria evaluation are discussed. We have shown that multicriteria methods provide a flexible way of dealing with qualitative environmental effects of decisions. However, this does not mean that multicriteria evaluation is a panacea which can be used in all circumstances without difficulties. It has its own problems, and some of these problems have also been illustrated. Finally, a comparison of the key characteristics of cost-benefit analysis and multicriteria evaluation has been carried out on the base of 13 comparison criteria, that is: rationality assumed, mathematical axiomatization, economic axiomatization, problem structuring, analytical cost, alternatives taken into account, evaluation criteria, preference system, democratic basis, aggregation procedures, comprehensiveness, empirical applicability, transparency and sustainability.

To choose between CBA and MCDA is not a trivial point, we could say that it is a meta-multicriteria problem. Van Pelt [1993] maintains that CBA is more attractive than MCDA from a methodological point of view, while MCDA is preferable from an empirical point of view. However, we have shown that from a pure methodological point of view, CBA has the advantage of being consistent with Neo-Classical economics; if other economic paradigms are considered this is not an advantage anymore. Furthermore, some assumptions underlying the rationality of CBA (e.g., utility maximisation, transitivity of indifference and preference relations) can also be disputable.

One should note that regarding environmental problems, CBA and MCDA can be considered as competitive methods only if all environmental consequences of decisions can be correctly transformed in monetary values; but we have seen in Chapter 3 that this is very difficult. Thus we can say that, given the presence of unpriced environmental impacts, often multicriteria evaluation is the only possible approach (if also fuzzy uncertainty is considered, the use of MCDA is even more advisable).

However, we do not conjecture that conventional monetary evaluation tools are to be discarded. CBA remains useful as one of the possible inputs to decision making, as long as policy makers bear its limitations in mind.

The second and the third objective of this study have been tackled in Part B. Our overview of multicriteria evaluation methods in a fuzzy environment, shows the following open problems:

- most of them are limited to the use of triangular fuzzy numbers;
- the shape of the membership function is not taken into consideration or only a part of it is used (leading to a loss of information);
- a general problem is the one of the "sensitivity" (degree of discrimination) of the solutions;
- all these methods are utility based models (thus based on the complete transitive comparability axiom) aimed at finding a single "optimal" solution;
- most of these methods are limited to the use of only fuzzy information.

As one can see a key issue in multicriteria evaluation in a fuzzy environment is how to compare fuzzy sets. Chapter 6 deals with the problem of comparison of fuzzy sets; a new approach based on a semantic distance using areas instead of traditional intersections has been presented. This semantic distance presents the following main characteristics:

(1) the absolute value metric is a particular case of this type of distance;
(2) the expected value is obtained as a representation of fuzzy sets only when their intersection is empty;
(3) the comparison between a fuzzy number and a crisp number is equal to the difference between the expected value of the fuzzy number and the value of the crisp number considered;
(4) when the intersection between two fuzzy sets is not empty, their distance is greater than the difference between their expected values;
(5) in case of fuzzy information represented by L-R fuzzy numbers, when their intersection is empty, their distance is equal to the distance between their middle values only when they are symmetric; otherwise, their expected values are obtained;
(6) by means of this semantic distance, the problem of the use of only one side of the membership functions, common to most of the traditional fuzzy multicriteria methods, is overcome;
(7) a theoretical justification of the importance of the area of the intersection in the comparison between two fuzzy sets can be drawn.

Since the complexity of environmental problems is high, there is a clear need for models offering a comprehensible and operational representation of a real-world environmental system. Qualitative aspects are hard to deal with in traditional models and therefore there is a clear need for methods that are able to take into account information of a "mixed" type (both qualitative and quantitative measurements). A problem, related to all multicriteria methods that try to take mixed information into account is the problem of equivalence of the procedures used in standardising the various evaluations of the performance of alternatives according to different criteria.

Given a "consistent family" of evaluation criteria $G=\{g_m\}$, m=1,2,..., M, and a finite set $A=\{a_n\}$, n=1, 2,..., N of potential alternatives (actions), a new multicriteria method called NAIADE (**N**ovel **A**pproach to **I**mprecise **A**ssessment and **D**ecision **E**nvironments) has been presented. It is a discrete multicriteria method whose impact (or evaluation) matrix may include either crisp, stochastic or fuzzy measurements of the performance of an alternative a_n with respect to a judgement criterion g_m.

From an empirical point of view, this model is particularly suitable for economic-ecological modelling incorporating various degrees of precision of the variables taken into consideration. From a methodological point of view, two main issues are then faced:

- the problem of equivalence of the procedures used in order to standardise the various evaluations (of a mixed type) of the performance of alternatives according to different criteria;
- the problem of comparison of fuzzy numbers typical of all fuzzy multicriteria methods.

Since in a fuzzy context, any attempt to reach a high degree of precision on the results tends to be somewhat artificial, a pairwise linguistic evaluation of alternatives is used. This is done by means of the notion of fuzzy relations (based on the new semantic distance) and linguistic quantifiers. In the aggregation process, particular attention is paid to the problem of diversity of the single evaluations, while the entropy concept is used as a measure of the associated "fuzziness". Such linguistic evaluations can be used in different ways according to the decision environment at hand.

In short, the main properties of the NAIADE method can be synthesised as follows:

(1) communication with the decision-maker is required to elicitate different relevant parameters, thus a constructive decision aid framework is implied;

(2) the method is based on some aspects of the partial comparability axiom, in particular, a pairwise comparison between alternatives is carried out, and incomparability relations are allowed;

(3) intensity of preference is taken into account, this implies that a certain degree of compensation between criteria is allowed, given the characteristics of the method it may be classified among partial compensatory methods;

(4) for the indifference relation no transitivity is implied, the preference relation is max-min transitive;

(5) a total or partial order of feasible alternatives is supplied, thus the γ problem formulation is tackled. It has to be noted that the final ranking is a function of all the alternatives considered, this implies that if a dominated or a dominating action is introduced, the ranking may change; moreover if the best action is eliminated, the ranking of the other alternatives may also change, thus NAIADE does not respect the independence of irrelevant alternatives axiom.

Two critical factors in determining the results provided by the NAIADE method seem to be the parameter α used in the aggregation process, and the aggregation operator used for the approximate reasoning operations. Therefore, in applying NAIADE, the analysis of the robustness of the results obtained according to these two factors may be useful. An example of sensitivity analysis in the framework of NAIADE has been presented. In the light of the results obtained, the following conclusions can be drawn. The pairwise linguistic evaluations offer useful complementary information to the final ranking. In fact, the former give indications of the relative credibility degree of preference, while the latter is only ordinal in nature. This is particular interesting for the evaluations on the indifference relations, since these are not considered in the ϕ^+ and ϕ^- indices. Moreover, one should note that in the ϕ^+ and ϕ^- indices the relationships among all the actions are considered, while the linguistic evaluations consider only a pair of actions.

When high values of α are considered, it is quite difficult to arrive at a final pairwise linguistic evaluation. This is mainly due to the presence of the threshold $\omega=0.5$ in the membership function defining the linguistic quantifier "most"; however by means of the Zimmermann-Zysno γ-operator, a higher degree of discrimination is possible. This is due to the fact that the γ-operator allows a certain degree of compensation between the values of the fuzzy relations and of the variable C, while the min operator is completely non-interactive. The presence of the threshold is also the reason why by means of the ϕ^+ and ϕ^- indices, it is still possible to rank actions for high values of α; however, one should note that in these situations only a few criteria with big intensities of preference could really play a role in the ranking process. In any case, we think that the use of high values of α may give some useful information above all when phenomena of rank reversal occur. In fact, in these cases the presence of rank reversals means that "discordant" criteria with big intensities of preference exist, thus information similar in spirit, to the one supplied by the veto threshold in ELECTRE methods could be obtained.

Finally, it should be noted that in the example shown the choice of the min operator or of the γ-operator does not seem to have any influence on the final rankings. This of course does not imply that this result is always true; furthermore, other operators exist that could supply different results, however in general, the choice of a linguistic operator is a very difficult problem.

Since in environmental and resource management and policy aiming at an ecologically sustainable development many conflicting issues and interests emerge, particular attention has to be given to the problem of different values and goals of different groups in society. Equity and conflicting values in multicriteria decision aid are traditionally introduced in two different ways:

(1)   by weighting the different criteria, but often in public decision making a single point-value solution (e.g. weights) tends to lead to deadlocks in a decision process because it imposes too many rigid conditions for a compromise;

(2)   by taking into consideration a set of ethical evaluation criteria. A weak point of this approach is that it could lead to an excessive number of evaluation criteria, furthermore, to identify ethical criteria may be not an easy task.

We have dealt with a third possibility i.e. the use of conflict analysis procedures to be integrated with multicriteria evaluation in order to allow policy-makers to seek for "defendable" decisions that could reduce the degree of conflict (in order to reach a certain degree of consensus) or that could have a higher degree of equity on different income groups. Starting with a matrix showing the impacts of different courses of action on each different interest/income group, a fuzzy clustering procedure indicating the groups whose interests are closer in comparison with the other ones is used. Therefore, finally a compromise solution taking into account all the three conflictual values of economics (efficiency, equity and sustainability) can in principle be isolated.

The empirical relevance of the developed mathematical procedures has been tested by means of a real-world environmental management problem (in the area of the Delta of the river Po in Italy). From this case study the following main conclusions can be drawn:

240

(1) multicriteria evaluation can help in finding a compromise solution between
conflictual ecological and economic objectives;

(2) the use of fuzzy sets can be a very useful tool in modelling environmental
management problems characterised by deep uncertainties and approximate
evaluations;

(3) the mathematical procedures developed in this study, can be an efficient tool
to deal with efficiency aspects, equity aspects and economy-environment
interactions of an environmental problem;

(4) the institution of a natural park may have positive consequences also from the
economic point of view, if economic activities more compatible with the natural
environment are carried out.

12.3   Main Limitations of the Present Study and Future Directions of Research

The debate between conventional economics and ecological economics may
give fruitful results. One should note that conventional economics can arrive at
elegant and clear-cut conclusions, ecological economics however shows that
assumptions on technical progress and substitution need to be revised in the light
of biophysical and ecological laws. Our personal position is more in favour of
ecological economics, but only years of future research can try to solve this
problem.

To establish under which conditions it is better to apply multicriteria evaluation
or cost-benefit analysis is not an easy task. In this study, we have tried to isolate
some criteria which may help the solution of this meta-multicriteria problem;
however the scientific discussion is still open. The main message we would like to
transmit is that the most important point is to try to tackle this problem on
scientific grounds and not on dogmatic ones!

One should note that when an attempt is made to model a real world
situation, the presence of a certain subjective component appears to be an
inevitable phenomenon. In general, this is a desirable feature, in fact when a
model without any creative, personal or subjective influence of a model designer
is used, this is inevitably characterised by a certain rigidity which prevents it
adhering completely to the situation modelled. This could make it necessary to
"force reality" because in the end the tendency will be to make reality fit the
model. The use of models with characteristics of subjectivity or of subjectivism,
depends in the latter analysis *on the ability and ethical behaviour of the
researcher constructing the model.* The analyst is generally subject to pressures
of politicians or stake holders who want to influence the outcome of the

evaluation process. It is important to remember this above all, when MCDA or CBA methods are used to "justify" or "defend" political decisions.

With no doubt a more rigorous axiomatic justification of some properties of the NAIADE method is required. For example, under which conditions rank reversals occur? Is it possible to establish the consequences of the introduction of the entropy concept in the "leaving and entering flows method"? Are there relationships between the degree of entropy and the degree of compensation allowed in the aggregation process? From a theoretical point of view, all these are very interesting questions to be answered, various developments in the field of MCDA theory could be reached by following these lines of research.

The variety of operators for the aggregation of fuzzy sets might make it difficult to decide which one to use; in fact, such a variety of operators makes the theory more flexible but at the same time more confusing in model formulations. In general, it is not only important that the operators satisfy certain axioms or have certain formal qualities from a mathematical point of view, but the operators must also be appropriate models of real system behaviour, which can normally only be proven by empirical testing.

A weak element in the fuzzy conflict analysis procedure developed here is the lack of strategic considerations leading to new coalitions or alliances. In fact, the clustering algorithm only indicates the groups whose interests are closer in comparison to the other ones; game-theoretic elements such as the notion of "power" need to be introduced. Furthermore, attaching to each interest group the same weight can be an oversimplification of a real-world situation. Future research in this direction could very much improve the treatment of equity issues in evaluation theory.

From an empirical point of view, the NAIADE method seems to be very suitable for environmental management applications. In this type of problems the presence of qualitative information is very common. The "boscone della Mesola" case study shows that fuzzy set theory can provide useful tools in modelling situations characterised by approximate evaluations; thus good multicriteria methods able to deal with this kind of information are a sensitive step for improving environmental management methods. Thanks to the use of the NAIADE method and the fuzzy conflict resolution procedure, in the "boscone della Mesola" application, it was possible to operationalize allocative, distributional and scale aspects of the problem, this seems a meaningful undertaking. However, the empirical behaviour of the mathematical procedures developed in this study needs to be checked in future applications of various nature. Only after many applications we will know if further improvements are needed.

REFERENCES AND BIBLIOGRAPHY

Ackoff R.L. (1978) - The art of problem solving, Wiley, New York.

Aguilera Klink F., Alcántara V. (eds.), (1994) - De la economía ambiental a la economía ecológica, Icaria, Barcelona (in Spanish).

Anderberg M. R. (1973) - Cluster analysis for applications, Academic Press, New York.

Anderson G., Bishop R. (1986) - The valuation problem, in Bromley D. (ed.) - Natural resource economics, Kluwer-Nijhoff, Boston, pp. 27-48.

Anderson N.H. (1981) - Foundations of information integration theory, New York, Academic Press.

Anderson N.H. (1982) - Methods of information integration theory, New York, Academic Press.

Archibugi F., Nijkamp P. (eds.), (1990) - Economy and Ecology: towards sustainable development, Kluwer, Dordrecht.

Arrow K.J. (1951) - Social choice and individual values, Wiley, New York.

Arrow K.J., Debreu G. (1954) - Existence of an equilibrium for a competitive economy, Econometrica, 2 (3).

Arrow K.J., Chenery H.B., Minhas B.S., Solow R.M. (1961) - Capital-labour substitution and economic efficiency, The Review of Economics and Statistics, vol. 43(3), pp. 225-250.

Arrow K.J. (1966) - Discounting and public investment criteria, in Kneese A.V., Smith S.L. (eds.) - Water research, pp. 13-32, John Hopkins, Baltimore.

Arrow K.J., Fisher A.C. (1974) - Preservation, uncertainty and irreversibility, Q.J. Econ., 88, pp. 312-319.

Arrow K.J., Raynaud H. (1986) - Social choice and multicriterion decision making, M.I.T. Press,(USA).

Axelrod R. (1970) - Conflict of interest, Chicago.

Ayres R.U., Kneese A.V. (1969) - Production, consumption and externalities, American Economic Review, 59, pp. 282-297.

Ayres R.U., Kneese A.V. (1990) - Externalities: economics and thermodynamics, in Archibugi F., Nijkamp P. (eds.) - Economy and Ecology: towards sustainable development, Kluwer, Dordrecht, pp.89-118.

Baas M.S., Kwakernaak H. (1977) - Rating and ranking of multiple aspect alternatives using fuzzy sets, Automatica, 13, pp. 47-58.

Baldwin J.F., Guild N.C.F. (1979) - Comparison of fuzzy sets on the same decision space, Fuzzy Sets and Systems, 2, pp. 213-232.

Bana e Costa C.A. (ed.) (1990a)- Readings in multiple criteria decision aid, Springer-Verlag, Berlin,.

Bana e Costa C.A. (1990b) - An additive value function technique with a fuzzy outranking relation for dealing with poor intercriteria preference information, in Bana e Costa C.A. (ed.)- Readings in multiple criteria decision aid, Springer-Verlag, Berlin, pp. 351-382.

Bana e Costa C.A. (1993) - Les problématiques dans le cadre de l'activité d'aide à la décision, Cahier du LAMSADE, n. 80, Paris

Banzhaf J.F. (1965) - Weighted voting doesn't work: a mathematical analysis, Rutgers Law Review, 19 pp. 317-343.

Barbier E.B. (1987) - The concept of sustainable economic development, Environmental Conservation, 14(2), pp. 101-110.

Barbier E.B. (1989) - Economics, natural resource scarcity and development: conventional and alternative views, Earthscan, London.

Barbier E.B., Markandya A, (1990) - The conditions for achieving environmentally sustainable growth, European Economic Review, 34, pp. 659-669.

Baumol W.J. (1968) - On the social rate of discount, American Economic Review, vol. 58, pp. 788- 802.

Baumol W.J. and Oates W.E. (1975) -The theory of environmental policy, Prentice-Hall, Englewood Cliffs,.

Bell D.E., Keeney R.L. and Raiffa H. (eds.) (1977) - Conflicting objectives in decision- International series on applied systems analysis- J. Wiley and Sons, New York.

Bell D.E., Raiffa H. and Tversky A.(eds.) (1988) - Decision making: descriptive, normative and prescriptive interactions, Cambridge University Press.

244

Bellman R.E., Zadeh L.A. (1970) - Decision-making in a fuzzy environment, <u>Management Science,</u> 17, pp. 141-164.

Bellman R.E., Giertz M. (1973,) - On the analytic formalism of the theory of fuzzy sets,<u>Information Sciences,</u> 6, pp. 149-156.

Belton V., Vickers S. (1990) - Use of a simple multi-attribute value function incorporating visual interactive sensitivity analysis for multiple criteria decision making, in Bana e Costa C.A. (ed.)- <u>Readings in multiple criteria decision aid,</u> Springer-Verlag, Berlin, pp. 319-334.

van den Bergh J.C.J.M., Nijkamp P. (1991) - Operationalizing sustainable development: dynamic ecological economic models, <u>Ecological Economics,</u> 4, pp.11-23.

van den Bergh J.C.J.M. (1991) - <u>Dynamic models for sustainable development,</u> Ph.D dissertation, Tinbergen Institute, Amsterdam.

van den Bergh J.C.J.M., van der Straaten J. (1994) - <u>Toward sustainable development,</u> Island Press, Washington.

Bezdek J.C., Spillman B., Spillman R. (1978) - A fuzzy relation space for group decision theory, <u>Fuzzy Sets and Systems,</u> 1, pp. 255-268.

Bezdek J.C., Spillman B., Spillman R. (1979) - Fuzzy relation spaces for group decision theory: An application, <u>Fuzzy Sets and Systems,</u> 2, pp. 5-14 .

Bezdek J.C. (1980) - <u>Pattern recognition with fuzzy objective functions algorithms,</u> Plenum, New York and London.

Bogetoft P., Pruzan P. (1991) - <u>Planning with multiple criteria,</u> North-Holland, Amsterdam.

Bouyssou D. (1986) - Some remarks on the notion of compensation in MCDM, <u>European Journal of Operational Research</u> 26 pp.150-160.

Bouyssou D. (1990) - Bulding criteria: a prerequisite for MCDA, in Bana e Costa C.A. (ed.)- <u>Readings in multiple criteria decision aid,</u> Springer-Verlag, Berlin, pp. 58-80.

Bouyssou D. ,Perny P. (1990) - Ranking methods for valued preference relations: a characterisation of a method based on leaving and entering flows, <u>Cahier du LAMSADE,</u> n. 101, Paris.

Bonissone P.P. (1982) - A fuzzy sets based linguistic approach: theory and applications, in M.M. Gupta and E. Sanchez (eds.)-<u>Approximate reasoning in decision analysis,</u> North Holland, Amsterdam, pp.329-338.

Bortolan G., Degani R. (1985) - Ranking fuzzy subsets, <u>Fuzzy Sets and Systems,</u> 15, pp. 1-19.

Braat L.C., van Lierop W.F.J. (eds.) (1987) - <u>Economic-ecological modeling,</u> North-Holland, Amsterdam.

Braat L.C. (1992) - <u>Sustainable multiple use of forest ecosystems: A simulation and evaluation model,</u> Ph.D. dissertation, Free University, Amsterdam.

Brans J.P., Mareschal B. and Vincke Ph. (1986) - How to select and how to rank projects. The PROMETHEE method, <u>European Journal of Operational Research,</u> vol.24, pp. 228-238.

Bresso M. (1993) - <u>Per un' economia ecologica,</u> La Nuova Italia Scientifica, Roma.

Bojö J., Mäler K.G. and Unemo L. (1990) - <u>Environment and development: an economic approach,</u> Kluwer Academic Publishers, Dordrecht.

Bojö J. (1991) - Economic analysis of environmental impacts, in Folke C., Kaberger T.-<u>Linking the natural environment and the economy: Essays from the Eco-Eco Group,</u> Kluwer, Dordrecht, pp.43-60.

Boulding K.E. (1966) - The economics of the coming spaceship earth, in Jarret H. (ed.)- <u>Environmental quality in a growing economy,</u> John Hopkins University Press, Baltimore, MD, pp. 3-14.

Brealey R., Myers S. (1985) - <u>Principles of corporate finance,</u> McGraw Hill, New York.

Cabeza Gutés M. (1995) - On the concept of weak sustainability, <u>Working Paper 284.95,</u> Universitat Autonome de Barcelona.

Capocelli R.M., De Luca A. (1973) - Fuzzy sets and decision theory, <u>Information and Control,</u> 23, pp. 446-473.

Cavalli S., Moschini R., Saini R. (1990) - <u>I parchi nazionali in Italia,</u> Unione delle Provincie Italiane, Roma.

Cavallo R. (1979) - Science, systems methodology, and the "interplay between nature and ourselves", in <u>Systems methodology in social science research,</u> Kluwer-Nijhoff, Boston, pp. 3-17.

Chalmers A. (1978) - <u>What is this thing called science?,</u> Open University Press.

Clark C.W. (1990) - <u>Mathematical Bioeconomics: the optimal management of renewable resources,</u> 2-nd ed., Wiley-Interscience, New York.

Clark C.W. (1991) - Economic biases against sustainable development, in Costanza R. (ed.)- Ecological Economics: the science and management of sustainability, Columbia University Press, New York, pp. 319- 320.

Coase R. (1960) - The problem of social cost, Journ. of Law and Economics, 3, pp. 1-44.

Coleman J.S. (1977) - Control of collectivities and the power of a collectivity to act, in Liebermann B. (ed.)- Social choice, New York, Academic Press, pp. 35-50.

Common M., Perrings C. (1992) - Towards an ecological economics of sustainability, Ecological Economics 6, pp. 7- 34.

Coombs C.H. (1958) - On the use of inconsistency of preferences in psychological measurement, Journal of Experimental Psychology, vol. 55, pp. 1-7.

Costanza R. (1987) -Social traps and environmental policy, BioScience, 37, pp. 407-412.

Costanza R., Perrings C. (1990) - A flexible assurance bonding system for improved environmental management, Ecological Economics 2, pp. 57-75.

Costanza R. (ed.) (1991) - Ecological Economics: the science and management of sustainability, Columbia University Press, New York.

Costanza R., Daly H.E., and Barthlomew J.A. (1991) - Goals,agenda and policy recommendations for ecological economics, in Costanza R. (ed.)- Ecological Economics: the science and management of sustainability, Columbia University Press, New York, pp. 1- 20.

Daly H.E. (1977) - Steady-State economics, Freeman, San Francisco, CA.

Daly H.E. (1979) - Entropy, growth and the political economy of scarcity, in Smith V.K. (ed.)- Scarcity and growth reconsidered, The Johns Hopkins University Press, Baltimore, pp. 67- 94.

Daly H.E., Cobb J.J. (1990) -For the common good: redirecting the economy toward community, the environment and a sustainable future, Beacon Press, Boston.

Daly H.E. (1991a) - Steady-State economics (2-nd edition), Island Press, Washington, DC.

Daly H.E. (1991b) - Elements of environmental macroeconomics, in Costanza R. (ed.)- Ecological Economics: the science and management of sustainability, Columbia University Press, New York, pp. 32-46.

Dasgupta A.K., Pearce D.W. (1972) - Cost-benefit analysis, MacMillan, London.

Dasgupta P., Heal D. (1979) - Economic theory and exhaustible resources, Cambridge University Press, London.

Davidson D. (1990) - Paradoxes of irrationality, in Moser P. K. (ed.)- Rationality in action, Cambridge University Press, pp. 449-464.

D'Avignon G., Vincke Ph. (1988) - An outranking method under uncertainty, European Journal of Operational Research 36, pp. 311-321.

De Campos L.M., Moral S. (1990) - Fuzzy measures with different levels of granularity, in Janko W. H., Roubens M. and Zimmermann H. J.- Progress in fuzzy sets and systems, Kluwer Academic Publishers, Dordrecht, pp. 134- 146.

DeLuca A., Termini S. (1972) - A definition of a non-probabilistic entropy in a setting of fuzzy sets theory, Information and Control, 20, pp. 301-312.

De Swaan A. (1973) - Coalition theories and cabinet formation, North-Holland, Amsterdam.

Dubois D., Prade H. (1980) - Fuzzy sets and systems: theory and applications, Academic Press, New York and London.

Dubois D., Prade H. (1986) - Fuzzy sets and statistical data, European Journal of Operational Research, 25, pp. 345-356.

Dubois D., Prade H. (1989) - Fuzzy sets, probability and measurement, European Journal of Operational Research, 40, pp. 135-154.

Enta Y. (1982) - Fuzzy decision theory, in Yager R.R. (ed.)- Fuzzy sets and possibility theory, Academic Publisher, New York, pp. 439-449.

Fandel G., Matarazzo B. and Spronk J. (eds.) (1983) - Multiple criteria decision methods and applications, Springer-Verlag, Berlin.

Faucheux S., Pilet G. (1994) - Energy metrics: on various valuation properties of energy, in Pethig R. (ed.) Valuing the environment. Methodological and measurement issues, Kluwer, Dordrecht.

Faucheux S., O'Connor M. (1994) - Conditions and indicators of sustainable development in an ecological-economic framework, paper presentato al "Congress of the Pacific regional science association", Cuernvaca, Mexico.

Faucheux S., Froger G., Munda G. (1994)- Des outils d' aide à la decision pour la multidimensionalité systémique: une application au développement durable, Revue Internationale de Systémique, No. 15.

Faucheux S., O'Connor M., van der Straaten J. - (1995)- Sustainable development: analysis and public policy, Kluwer, Dordrecht.

Fedrizzi M., Kacprzyk J. (1988) - On measuring consensus in the setting of fuzzy preference relations, in Kacprzyk J., Roubens M.- Nonconventional preference relations in decision making, Springer-Verlag, Berlin, pp. 129, 141.

Feyerabend P. (1975) - Against method, New Left Books, London.

Folke C., Kaberger T. (eds.) (1991a) - Linking the natural environment and the economy: Essays from the Eco-Eco Group, Kluwer, Dordrecht.

Folke C., Kaberger T. (1991b) - Recent trends in linking the natural environment and the economy, in Folke C., Kaberger T. (eds.)- Linking the natural environment and the economy: Essays from the Eco-Eco Group, Kluwer, Dordrecht, pp. 273- 300.

Folke C. (1991) - Socio-economic dependence on the life-supporting environment, in Folke C., Kaberger T. (eds.)- Linking the natural environment and the economy: Essays from the Eco-Eco Group, Kluwer, Dordrecht, pp. 77-94.

Freksa C. (1982) - Linguistic description of human judgements in expert systems and in the "soft sciences", in Gupta M.M. and Sanchez E. (eds.)- Approximate reasoning in decision analysis, North-Holland, Amsterdam, pp. 297-305.

French S. (1984) - Fuzzy decision analysis: some criticism, in Zimmermann H.J., Zadeh L.A. and Gaines B.R.- Fuzzy sets and decision analysis, TIMS Studies in the Management Sciences, 20, pp. 29-44.

French S. (1986) - Decision theory, Ellis Horwood.

French S. (1992) - Uncertainty and imprecision: modelling and analysis, paper presented at the 4-th International Meeting of Statistics in the Basque Country.

Funtowicz S.O., Munda G., Paruccini M. (1990) - The aggregation of environmental data using multicriteria methods, Environmetrics, Vol. 1(4), pp. 353-368.

Funtowicz S.O., Ravetz J.R. (1990) - Uncertainty and quality in science for policy, Kluwer Academic Publishers, Dordrecht.

Funtowicz S.O., Ravetz J. R. (1991) - A new scientific methodology for global environmental issues, in R. Costanza (ed.)- Ecological Economics, New York, Columbia, pp. 137-152.

Funtowicz S.O., Ravetz J. R. (1994a) - The worth of a songbird: ecological economics as a post-normal science, Ecological Economics, 10, pp. 197-207.

Funtowicz S.O., Ravetz J. R. (1994b) - Emergent complex systems, Futures, 26(6), pp. 568-582.

Galbraith J.K. (1959) - The affluent society, Houghton-Mifflin, New York.

Gambarelli G. (1983) - Common behaviour of power indices, International Journal of Game Theory, 12, 4, pp. 237-244.

Gamson W.A. (1962) - Coalition formation at presidential nominating conventions, American Journal of Sociology, 68, pp. 157-171.

Gamson W.A. (1968) - International Encyclopaedia of the social sciences, entry Coalitions.

Geoffrion A. M. , Dyer J. S. , Feinberg A. (1972) - An interactive approach for multicriterion optimization with an application to the operation of an academic department", Management Science, n.23, pp. 357-368.

Georgescu-Roegen N. (1971) -The entropy law and the economic process, Harward University Press, Cambridge MA.

Georgescu-Roegen N. (1979) - Comments on the papers by Daly and Stiglitz, in Smith V.K. (ed.)- Scarcity and growth reconsidered, The Johns Hopkins University Press, Baltimore, pp. 95-105.

Georgescu-Roegen N. (1993) - Thermodynamics and we, the humans, in Dragan J.C., Seifert E.K., Demetrescu M.C. (eds.) - Entropy and Bioeconomics, NAGARD, Milano.

Giarlotta A. (1990) - Multicriteria compensability analysis ranking totally the alternatives based on the employment of a non-symmetric information axiom (CARTESIA), Annali della Facoltà di Economia e Commercio, Catania (in Italian).

Ginzberg M. J., Stohr E. A., (1982) - Decision support systems: issues and perspectives- in Ginzberg M. J., Reitman W. and Stohr E. A. (eds.)-Decision support systems, North-Holland Publ. Co., Amsterdam, pp. 9-32.

Gonzalez Muñoz A., Vila M.A. (1990) - Dominance and indifference relations on fuzzy numbers, in Janko W. H., Roubens M. and Zimmermann H. J.- Progress in fuzzy sets and systems, Kluwer Academic Publishers, Dordrecht, pp. 75-89.

Goodland R., Ledec G. (1987) -Neoclassical economics and principles of sustainable development, Ecological Modeling, vol.38, pp. 19-46.

Gowdy J.M. (1991) -Bioeconomics and post Keynesian economics: a search for common ground, Ecological Economics, 3 , pp. 77-87.

Greco S. (1992) - L'approccio delle regole di decisione intercriteriali: il metodo IDRA, Atti del Sedicesimo Convegno A.M.A.S.E.S., Treviso, pp. 431-444 (in Italian).

de Groot R.S. (1992) - Functions of nature, Wolters-Noordhoff, Groningen.

Guimaraes Pereira A., Munda G. and Paruccini M. (1994) - Generating alternatives for siting retail and service facilities using genetic algorithms and multiple criteria decision techniques, Journal of Retailing and Consumer Services, vol. 1, N. 1, pp. 40 - 47.

Gupta M.M., Saridis G. N. and Gaines B.R. (eds.) (1977) - Fuzzy automata and decision processes, North-Holland, Amsterdam.

Gupta M.M. and Sanchez E. (eds.) (1982) - Approximate reasoning in decision analysis, North-Holland, Amsterdam.

Hafkamp W. (1984) - Economic-environmental modeling in a national-regional system, North-Holland, Amsterdam.

Hanley N. (1988) - Using contingent valuation to value environmental improvements, Applied Economics, 20 (4), pp. 541-550.

Hanley N. (1992) - Are there environmental limits to cost benefit analysis?, Environmental and Resource Economics, 2, pp. 33-59.

Hanley N., Spash C. (1993) - Cost-benefit analysis and the environment, Edward Elgar, Aldershot.

Harcourt G.C. (1972) - Some Cambridge controversies in the theory of capital, Cambridge University Press, Cambridge, UK.

Hardin G. (1968) - The tragedy of the Commons, Science, 162, pp. 1243-1248.

Harker P.T. (1989) - The art and science of decision making: the analytic hierarchy process, in Golden B.L., Wasil E.A., Harker P.T. (eds.) - The analytic hierarchy process, Springer-Verlag, Berlin, pp. 3-36.

Hartigan J. (1975) - Clustering algorithms, John Wiley and Sons, New York.

Hartwick J.M. (1977) - Intergenerational equity and the investing of rents from exhaustible resources, American Economic Review, vol.67(5), pp. 972-974.

Hartwick J.M. (1978) - Substitution among exhaustible resources and inter-generational equity, Review of Economic Studies, vol. 45, pp. 347-354.

Heise D.R. (1969) - The semantic differential and attitude research, in Summers G.F. (ed.)-Attitude measurement, Chicago, Rand McNally, pp. 8-20.

Helmers R.L.C.H. (1979) - Project planning and income distribution, Martinus Nijhoff, Boston.

Hersh H.M. and Caramazza A. (1976) - A fuzzy set approach to modifiers and vagueness in natural languages, Jour. of experimental psychology: General, 105, pp. 254-276.

Hersh H.M., Caramazza A. and Brownell H.H. (1979) - Effects of context on fuzzy membership functions, in Gupta M.M., Regade R.K. and Yager R.R. (eds.)- Advances in fuzzy set theory and applications, Amsterdam, North Holland, pp. 389-408.

Hesketh B. and Roche S. (1986) - Research into classification procedures at a youth prison, Australia and New Zeland Journal of Criminology, 19, pp. 42-52.

Hesketh B., Preyor R., Gleitzman M. and Hesketh T. (1988) - Practical applications and psychometric evaluation of a computerized fuzzy graphic rating scale, in Zétényi T.- Fuzzy sets in psychology, North-Holland, Amsterdam, pp. 425-454.

Hicks J.R. (1939) - The foundations of welfare economics, Economic Journal, vol. 49.

Higashi M., Klir G.J. (1982) - Measures of uncertainty and information based on possibility distributions, International Journal of General Systems, 9, pp. 43-58.

Hildenbrand W., Kirman A.P. (1991) - Equilibrium analysis, North-Holland, Amsterdam.

Hinloopen E., Nijkamp P. (1990) - Qualitative multiple criteria choice analysis, the dominant regime method- Quality and quantity 24, pp. 37-56.

Hirsch F. (1976) -Social limits to Growth, Harward University Press, Cambridge Mass.

Hirschleifer J. (1958) - On the theory of optimal investment decision, Journal of Political Economy, pp. 329-352.

Hirschleifer J. (1966) - Investment decision under uncertainty: application of the state-preference approach, Quarterly Journal of Economics, vol. 80, pp. 252 - 277.

Hilsdal E. (1985) - Are grades of membership probabilities?- paper presentes at the 1st IFSA Congress, Palma de Mallorca.

Hoevenagel R. (1994) - The contingent valuation method: scope and validity. Ph.D Thesis, Free University, Amsterdam.

Holler M.J. (ed.) (1984) - Coalitions and collective action, Physica-Verlag, Wuerzburg (Germany).

Hotelling H. (1931) - The economics of exhaustible resources, Journal of Political Economy, 39, pp. 137-175.

Hueting R. (1980) - New scarcity and economic growth: more welfare through less production?, North-Holland, Amsterdam.

IDROSER (1978) - Progetto di piano per la salvaguardia e l'utilizzo ottimale delle risorse idriche in Emilia Romagna, Volume V: La domanda idrica per usi naturalistici e ambientali, Bologna (in Italian).

IDROSER (1985a) - Progetto di fattibilità per la creazione del parco naturale del bosco di Mesola, Bologna (in Italian).

IDROSER (1985b) - Valutazione economica del riassetto ambientale dell'ecosistema del boscone della Mesola, Bologna (in Italian).

Isermann H. (1982) - Linear lexicographic optimization, OR-Spektrum 4, pp. 223-228.

Jain R. (1977) - Procedure for multi-aspect decision making using fuzzy sets, International Journal of Systems Science, 8, pp. 1-7.

Janko W. H., Roubens M. and Zimmermann H. J. (1990) - Progress in fuzzy sets and systems. Kluwer Academic Publishers, Dordrecht.

Janssen R. (1992) - Multiobjective decision support for environmental problems. Kluwer, Dordrecht.

Johansson P.O. (1987) - The economic theory and measurementof environmental benefits, Cambridge University Press, Cambridge.

Kaberger T. (1991) - Measuring instrumental value in energy terms, in Folke C., Kaberger T. (eds.) - Linking the natural environment and the economy: Essays from the Eco-Eco Group, Kluwer, Dordrecht.

Kaldor N. (1939) - Welfare comparison of economics and interpersonal comparisons of utility, Economic Journal, vol. 49.

Kaufmann A. (1975) - Introduction to the theory of fuzzy subsets, vol.1, Academic Press, New York.

Keeney R., Raiffa H. (1976) - Decision with multiple objectives: preferences and value trade-offs, Wiley, New York.

Keller W.J., Wansbeek T. (1983) - Multivariate methods for quantitative and qualitative data, Journal of Econometrics, vol. 22, pp. 91-111.

Keynes J.M. (1936) - The general theory of employment, interest and money, The MacMillan Press Ltd., London.

Klaassen G.A.J., Opschoor J. B. (1991) - Economics of sustainability or the sustainability of economics: different paradigms, in Ecological Economics, 4 , pp. 93-115.

Klir G. J.- An approach to general systems theory, Van Nortrand, New York, 1969.

Kmietowicz Z. W., Pearman A.D. (1981) - Decision theory and incomplete knowledge, Gower, Adershot.

Kneese A., Ayres R., d' Arge R. (1970) - Economics and the environment: a materials balance approach, Resource for the Future, Washington DC.

Korhonen P., Wallenius J. (1988) - A careful look at efficiency and utility in multiple criteria decision making: a tutorial, W.P., Helsinki School of Economics, F-197.

Kruskal J.B. (1964) - Multidimensional scaling by optimizing goodness of fit to a nonmetric hypothesis, Psychometrika, vol. 29, pp. 1-27.

Kuhn T.S. (1962) - The structure of scientific revolutions, University of Chicago Press, Chicago.

van Laarhoven P.J.M., Pedrycz W. (1983) - A fuzzy extension of Saaty's priority theory, Fuzzy Sets and Systems, 11, pp. 229-241.

Leclercq J.P. (1984) - Propositions d'extension de la notion de dominance en présence de relations d'ordre sur les pseudo-critéres: MELCHIOR, Revue Belge de Recherche Opérationelle, de Statistique et d' Informatique 24 (1), pp. 32-46

Leiserson M. (1966) - Coalitions in politics. Unpublished Ph.D. Dissertation,Yale.

Leung Y. (1988) - Spatial analysis and planning under imprecision. North Holland, Amsterdam.

Li D., Liu D. (1990) - A fuzzy PROLOG database system, John Wiley and Sons, New York.

Lichfield N. (1964) - Cost benefit analysis in plan evaluation, Town Planning Review, vol. 35, N. 2, pp. 160-169.

Lichfiel N. (1988) - Economics in urban conservation, Cambridge, Cambridge University Press.

Lichfield N. (1993) - Problems of valuation discussed, in Banister D., Button K. (eds.)- Transport, the environment and sustainable development, E & FN Spon, London, pp.205-211.

Lipsey R.G., Lancaster K. (1956) - The general theory of second-best, Review of Economic Studies, vol. 7.

Little I.M.D. (1950) - A critique of welfare economics, Oxford University Press, Oxford.

Lovelock J.E. (1988) - The ages of Gaia, Norton & Co., New York.

Luce R.D. (1956) - Semiorders and a theory of utility discrimination, Econometrica, 24, pp. 178-191.

Luce R.D. and Raiffa H. (1989) - Games and decisions, Dover, New York.

MacNelly J.A., Miller K.S. (eds.).(1984) - National parks, Conservation and development: the role of protected areas in sustaining society, Smithsonian Institution Press, Washington D.C.

Maier-Rigaud G. (1991) - Background to the conflict between economic and ecological ends, Ecological Economics 4 , pp. 83- 91.

Malthus T. (1798) - An essay on the principle of population, reprinted by Macmillan, London, 1909.

Mäler K.G. (1974) - Environmental economics, The Johns Hopkins Press, Baltimore.

Marchetti R. (1984) - Quadro analitico complessivo dei risultati delle indagini condotte negli anni 1977-1980 sul problema dell'eutrofizzazione nelle acque costiere dell'Emilia-Romagna: situazione e ipotesi di intervento, Regione Emilia-Romagna, Bologna (in Italian).

Martinez-Alier J. (1987) - Ecological Economics, Basil-Blackwell, Oxford.

Martinez-Alier J. (1994a) - De la economia ecológica al ecologismo popular, ICARIA, Barcelona (in Spanish).

Martinez-Alier J. (1994b) - Distributional issues in ecological economics, paper presented at the Third International Meeting of Ecological Economics, San Jose`, Costa Rica.

Martinez-Alier J. (1994c) - Distributional conflicts and international environmental policy on carbon dioxide emissions and agricultural biodiversity, in van den Bergh J.C.J.M., van der Straaten J. - Toward sustainable development, Island Press, Washington.

Martinez-Alier J., O'Connor M. (1995) - Ecological and economic distribution conflicts, to appear in Costanza R., Segura O. (eds.) - Getting down to Earth: practical applications of ecological economics, Island Press/ISEE, Washington D.C.

Marx K. (1867) - Capital, reprinted by Lawrence and Wishart, London, 1970.

Matarazzo B. (1981) - Sulla scelta degli investimenti privati, Univ. Catania, Ist. of Mathematics, Catania (in Italian).

Matarazzo B. (1986) - Multicriterion analysis of preferences by means of pairwise actions and criterion comparisons (MAPPAC)- Applied mathematics and computation, 18/2, pp. 119-141.

Matarazzo B. (1988) - Preference ranking global frequencies in multicriterion analysis (PRAGMA), European Journal of Operational Research 36/1 pp. 36-49.

Matarazzo B. (1991) -PCCA and k-dominance in MCDM, Belgian Journal of Operations Research, Statistics and Computer Science, Vol. 30, N. 3, pp. 56-69.

Mayumi K. (1991) - A critical appraisal of two entropy theoretical approaches to resources and environmental problems, and a search for an alternative, in Rossi C., Tiezzi E. (eds.)- Ecological physical chemistry, Elsevier, Amsterdam, pp. 109-130.

Mayumi K. (1992) - The new paradigm of Georgescu-Roegen and the tremendous speed of increase in entropy in the modern economic process, The Journal of Interdisciplinary Economics, vol. 4, pp. 101-129.

Mayumi K. (1993) - Georgescu-Roegen's "fourth law of thermodynamics", the modern energetic dogma, and ecological salvation, in Bonati L., Cosentino U., Lasagni M., Moro G., Pitea D., Schiraldi A. (eds.) - Trends in ecological physical chemistry, Elsevier, Amsterdam.

McKenzie G.W., Pearce I.F. (1982) - Welfare measurement: a synthesis, American Economic Review, 72, pp. 669-682.

Meadows D.H., Meadows D.L., Randers J. and Behrens W.W. (1972) - The limits to growth, Universe Books, New York.

Mill J.S. (1857) - Principles of political economy, Parker London.

Ministero dell'Agricoltura, Consorzio per il canale emiliano romagnolo (1990) - Po: Acqua Agricoltura Ambiente, Il Mulino, Bologna (in Italian).

Ministero dell' Ambiente (1991) - Studi per la pianificazione e il controllo del risanamento del bacino padano (MASTERPLAN), Roma (in Italian).

Mirowski P. (1989) - More heat than light, Cambridge University Press, Cambridge.

Mishan E.J. (1967) - The cost of economic growth, Staples Press, London.

Mishan E.J. (1971a) -Cost-benefit analysis, Allen and Unwin, London.
Mishan E.J. (1971b) -The postwar literature on externalities: an interpretative essay, Jour. of Economic Literature, 9 (1), pp.1-28.
Mishan E.J. (1977) - The economic growth debate: an assessment, George Allen & Unwin, London.
Miyamoto S. (1990) - Fuzzy sets in information retrieval and cluster analysis, Kluwer Academic Publishers, Dordrecht.
Morey E. R (1984) - Confuser surplus, American Economic Review, 74, pp. 163- 173.
Moser P. K. (ed.) (1990) - Rationality in action, Cambridge University Press.
Moulin H. (1988) - Axioms of co-operative decision making, Econometric Society Monographs, Cambridge University Press.
Mueller D. (1974) - Intergenerational justice and the social discount rate, Theory and Decision, vol. 5, pp. 263- 273.
Munda G. (1988) - Fuzzy sets: principi teorici ed applicazioni ai metodi MCDA, "Laurea" degree thesis, University of Catania (in Italian).
Munda G. (1989) -Multiple criteria decision aid: principi teorici ed applicazioni nel campo delle scienze regionali, in proceedings of the 10-th meeting of the Italian Regional Sciences Association, Rome, vol. 2, pp. 1225-1244 (in Italian).
Munda G., Nijkamp P. and Rietveld P. (1992) - Multicriteria evaluation and fuzzy set theory: applications in planning for sustainability, Serie Research Memoranda, Free University, Amsterdam.
Munda G. (1993) -Multiple criteria decision aid: some epistemological considerations, Journal of Multi-Criteria Decision Analysis, vol. 2, pp. 41-55.
Munda G., Nijkamp P. and Rietveld P. (1993a) - Information precision and multicriteria evaluation methods, in Giardina E., Williams A. (eds.)- Efficiency in the public sector, Edward Elgar, Aldershot, pp. 43-64.
Munda G., Nijkamp P. and Rietveld P. (1993b) - Comparison of fuzzy sets: a new semantic distance, Ricerca Operativa, vol. 23, N. 65, pp. 5-25.
Munda G., Nijkamp P. and Rietveld P. (1994a) - Qualitative multicriteria evaluation for environmental management, Ecological Economics, vol. 10, N. 2, pp. 97-112.
Munda G., Nijkamp P. and Rietveld P. (1994b)- Multicriteria evaluation in environmental management: why and how?, in Paruccini M. (ed.) - Applying multiple criteria aid for decision to environmental management, Kluwer, Dordrecht, pp. 1-22.
Munda G., Nijkamp P. and Rietveld P. (1994c)- Fuzzy multigroup conflict resolution for environmental management, in J. Weiss (ed.) - The economics of project appraisal and the environment, Edward Elgar, Aldeshot, pp. 161-183.
Munda G., Nijkamp P. and Rietveld P. (1995a) - Qualitative multicriteria methods for fuzzy evaluation problems, European Journal of Operational Research, N. 82, pp.79-97.
Munda G., Nijkamp P. and Rietveld P. (1995b) - Monetary and non-monetary evaluation methods in sustainable development planning, Economie Appliquée, tome XLVIII, N.2, pp.145-162.
Munda G., Nijkamp P. and Rietveld P. (1995c) - Environmental decision making: a comparison between cost-benefit analysis and multicriteria decision aid, to appear in Faucheux S., O'Connor M. and van der Straaten J. (eds.) - Sustainable development: analysis and public policy, Kluwer, Dordrecht.
Munro A., Hanley N. (1991) - Shadow projects and the stock of natural capital: a cautionary note, Discussion Paper NO. 91/7, Economics Department, University of Stirling.
Myrdal G. (1973) - Against the stream. Critical essays on economics, Random House, New York.
Myrdal G. (1978) - Institutional economics, J. Econ. Issues 12: pp. 771-783.
Nagel S.S., Mills M.K. (1991) - Systematic analysis in dispute resolution, Quorum Books, New York.
Nakayama H., Tanino T., Matsumoto K., Matsuo H., Inoue K., Sawaragi Y. (1979) - Methodology for group decision making with an application to assessment of residential environment, IEEE Trans. Systems Man & Cybernet, 9, pp. 477-485.
Naredo J.M. (1987) - La economía en evolución, Siglo XXI, Madrid.
Naredo J.M., Campos P. (1980) - Los balances energéticos de la agricultura española, Agricultura y Sociedad, N. 15.
Negoita C.V., Minoiu S., Stan E. (1976) - On considering imprecision in dynamic linear programming, Economic Computation and Economic Cybernetics Studies and Research, 3, pp. 83-95.
Neumann J. von, Morgenstern O. (1947) - Theory of games and economic behaviour, Princeton.

Nijkamp P., van Delft A. (1977) - Multicriteria analysis and regional decision-making, Martinus Nijhoff, Leiden.

Nijkamp P. (1979) - Multidimensional spatial data and decision analysis, Wiley, New York.

Nijkamp P. (1980) - Environmental policy analysis- New York, Wiley.

Nijkamp P., Leither H. and Wrigley N. (eds.) (1985) - Measuring the unmeasurable, NATO ASI Series, Martinus Nijhoff, Dordrecht.

Nijkamp P., Voogd M. (1985) - An informal introduction to multicriteria evaluation, in Fandel G., Matarazzo B., Spronk J. (eds.)- Multiple criteria decision methods and applications, Springer-Verlag, Berlin-Heidelberg, pp. 27-40.

Nijkamp P., Rouwendal J. (1987) - Time, discount rate and public decision-making, Serie Research Memoranda, Free University, Amsterdam.

Nijkamp P., Rietveld P. and Spronk J. (1988) - Open problems in the operationalization of multiple criteria decision methods. A brief survey.- Syst. Anal. Model. Simul. 5 4, pp.311-322.

Nijkamp, P., Rietveld P. and Voogd H. (1990) - Multicriteria evaluation in physical planning, North-Holland, Amsterdam.

Nijkamp P., van den Bergh J.C.J.M., and Soetman F.J. (1991) - Regional sustainable development and natural resource use, in Proceedings of the World Bank Annual Conference on development Economics.

Nijkamp P., Bithas K. (1992) - Sustainable development and monument conservation planning. A case study on Olympia, Serie Research Memoranda, Free University, Amsterdam.

Nordhaus W. D., Tobin J. (1972) - Is growth obsolete?, in National Bureau of Economic Research, Economic Growth, Fiftieth Anniversary Colloquium, vol. 5, New York.

Norese M.F. (1991) - A multiple criteria approach to complex situations, in Jackson et al. (eds.)- Systems thinking in Europe, Plenum Press, New York, pp. 361-369.

Norgaard R.B. (1988) - Sustainable development: a co-evolutionary view, Futures, vol. 20, No. 6, pp. 606-620.

Norgaard R. B. (1994) - Development Betrayed, Routledge, London.

Norwich A.M. and Turksen I.B. (1982) - The fundamental measurement of fuzziness, in Yager R.R. (ed.)- Fuzzy set and possibility theory, New York, Pergamon.

Norwich A.M. and Turksen I.B. (1984) - A model for the measurement of membership and the consequences of its empirical implementation, Fuzzy sets and systems, 12, pp. 1-25.

O'Connor M. (1991) - Entropy, structure, and organisational change, Ecological Economics, 3, pp. 95-122.

O'Connor M., Faucheux S., Froger G., Funtowicz S.O., Munda G. (1995) - Emergent complexity and procedural rationality: post-normal science for sustainability, to appear in Costanza R., Segura O. (eds.) - Getting down to Earth: practical applications of ecological economics, Island Press/ISEE, Washington D.C.

Oden G.C. (1977) - Integration of fuzzy logical information, Jour. of experimental psychology: Human perception and performance,3, pp. 565- 575.

Odum E.P. (1989) - Ecology and our endangered life-support systems, Sinuaer Associates, Sunderland, Massachussetts.

O'Neil J. (1993) - Ecology, policy and politics, Routledge, London.

Opschoor J.B. (ed.) (1992) - Environment, economy and sustainable development, Wolters-Noordhoff, Amsterdam.

Opschoor J.B., van der Straaten J. (1993) - Sustainable development: an institutional approach, Ecological Economics, 7, pp. 203-222.

Osgood C.E., Suci G.J. and Tannenbaum P.J. (1957) - The measurement of meaning, Urbana, III, Univ. of Illinois Press.

Ostanello A. (1990) - Action evaluation and action structuring: different decision aid situations reviewed through two actual cases, in Bana e Costa C.A. (ed.)- Readings in multiple criteria decision aid, Springer-Verlag, Berlin, pp. 36-57.

Owen G. (1968) - A note on the Shapley value, Management Science, pp. 731-732.

Owen G. (1971) - Political games, Naval Research Logistics Quarterly, 18, pp. 345-355.

Passet R. (1979) - L'economique et le vivent, Payot, Paris.

Pastijn H., Leysen J. (1989) - Constructing an outranking relation with ORESTE: Math. Computer Model., 12 (10/11), pp.1255-1268.

Patinkin D. (1963) - Demand curves and consumer's surplus, in Christ C.F. et al. (eds.)- Measurement in economics, Studies in mathematical economics and econometrics in memory of Yehula Greenfeld, Stanford University Press, Stanford.

Pearce D.W. (1971) - Cost-Benefit analysis, MacMillan, London.

Pearce D.W. (1976) - The limits of cost-benefit analysis as a guide to environmental policy, Kyklos, 29 (1), pp. 97-112.

Pearce D.W. (ed.) (1978) - The valuation of social cost, George Allen & Unwin, London,.

Pearce D.W., Markandya A. (1987) -Marginal opportunity cost as a planning concept in natural resource management, Ann. Reg. Sci., December.

Pearce D.W. (1983) - Ethics, irreversibility, future generations and the social rate of discount, International Journal of Environmental Studies, vol. 21, pp. 67-86.

Pearce D.W., Nash C.A. (1989) - The social appraisal of projects, MacMillan, London,.

Pearce D.W., Markandya A. and Barbier E. (1989) - Blueprint for a green economy, London, Earthscan.

Pearce D.W., Turner K.R. (1990) - Economics of natural resources and the environment, Harvester Wheatsheaf, New York.

Pearce D.W., Atkinson G.D. (1992) - Are national economies sustainable? Measuring sustainable development, CSERGE Working Paper GEC 92-11.

Pearce D.W., Atkinson G.D. (1993) - Capital theory and the measurement of sustainable development: an indicator of "weak" sustainability, Ecological Economics, vol. 8, pp. 103-108.

van Pelt M.J.F. (1993) - Sustainability-oriented project appraisal for developing countries, Ph.D. Dissertation, Wageningen Agricultural University, Wageningen.

Perny P., Roy B. (1992) - The use of fuzzy outranking relations in preference modelling, Fuzzy Sets and Systems, 49, pp. 33-53.

Pezzey J. (1989) - Economic analysis of sustainable growth and sustainable development, The World Bank, Environmental department Working Paper N. 15.

Pigou A.C. (1920) - The economics of welfare, Macmillan, London.

Premazzi G., Facchetti S., Freudenthal J. (1990) - A snapshot survey on the Po river, Commission of the European Communities, Joint Research Centre, Environment Institute, EUR 13037 EN, Ispra.

Quinet E. (1993) - Can we value the Environment?, in Banister D., Button K. (eds.)- Transport, the environment and sustainable development, E & FN Spon, London, pp. 191-204.

Ragade R.K. and Gupta M.M. (1977) - Fuzzy set theory: Introduction, in M.M. Gupta, G.N. Saridis and B.R. Gaines, Fuzzy automata and decision processes, North-Holland, Amsterdam,, pp. 105-131.

Ramik J., Rimanek J. (1985) - Inequality between fuzzy numbers and its use in fuzzy optimization, Fuzzy Sets and Systems 16, pp. 123-138.

Ricardo D. (1926) - Principles of political economy and taxation, Everyman, London.

Rietveld P. (1980) - Multiple objective decision methods and regional planning, North Holland, Amsterdam.

Rietveld P. (1984) -The use of qualitative information in macro-economic policy analysis, in Despontin M., Nijkamp P. and Spronk J.(eds.)- Macro-economic planning with conflicting goals, Springer-Verlag, Berlin, pp. 263-280.

Rietveld P. (1989) - Using ordinal information in decision making under uncertainty, Systems Analysis, Modeling, Simulation, vol. 6, pp. 659-672.

Riker W.H. (1962) - The theory of political coalitions, New Haven, London.

Rìos Insua D. (1990) - Sensitivity analysis in multi-objective decision making, Springer-Verlag, Berlin.

Riva A. (1983) - Adattamento e sviluppo dell'uomo, Pontificia Università Lateranense "ut unum sint", Roma (in Italian).

Roberts F. S. (1979) - Measurement theory with applications to decision making, utility and the social sciences, Addison-Wesley, London.

Rommelfanger H. (1989a) - Inequality relations in fuzzy constraints and its use in linear fuzzy optimization, in Verdegay J.L., Delgado M.- The interface between artificial intelligence and operations research in fuzzy environment, pp. 195-211.

Rommelfanger H. (1989b)- Interactive decision making in fuzzy linear optimization problem, European Journal of Operational Research, 41, pp. 210-217.

Roubens M., Vincke Ph. (1985) - Preference modelling, Springer- Verlag, Berlin.

Roubens M. (1990) - Inequality constraints between fuzzy numbers and their use in mathematical programming, in Slowinski R., Teghem J.- Stochastic versus fuzzy approaches to multiobjective mathematical programming under uncertainty, Kluwer Academic Publishers, Dordrecht, pp. 321- 330.

Roy B. (1968) - Clessement et choix en presence de points de vue multiple (la methode ELECTRE), R.I.R.O., 8, pp. 57-75.

Roy B., Bertier P. (1973) - La methode ELECTRE II. Une application ou media planning, in M.Ross (ed.), Operational research '72, North Holland, Amsterdam, pp. 291-302.

Roy B. (1977) - Partial preference analysis and decision aid: the fuzzy outranking relation concept, in Bell D.E., Keeney R.L. and Raiffa H. (eds.)- Conflicting objectives in decision- International series on applied systems analysis- J. Wiley and Sons, New York, pp.41-75.

Roy B. (1978) - ELECTRE III: Un algorithme de classement fondé sur une répresentation floue desprèferences en prèsence de critères multiples, Cahiers du Centre d' Etudes de Recherche Opèrationnelle Paris, 20 (1), pp. 3-24.

Roy B. (1985) - Méthodologie multicritere d' aide à la decision, Economica, Paris.

Roy B., Présent M., Sillhol D. (1986) - A programming method for determining which Paris metro stations should be renovated, European Journal of Operational Research, 24, pp. 318-334.

Roy B. (1989) - Main sources of inaccurate determination, uncertainty and imprecision in decision models, in Models and methods in multiple criteria decision making, edited by Colson G., De Bruyn Chr., Pergamon Press, Oxford, pp. 1245-1254.

Roy B. (1990a) - The outranking approach and the foundations of ELECTRE methods, in Bana e Costa C.A. (ed.)- Readings in multiple criteria decision aid, Springer-Verlag, Berlin, pp. 155-183.

Roy B. (1990b)- Decision aid and decision making, in Bana e Costa C.A. (ed.)- Readings in multiple criteria decision aid, Springer-Verlag, Berlin, pp. 17-35.

Roy B. (1990c) -Science de la decision ou science de l'aide a la decision? -Cahier n. 97, LAMSADE, Paris.

Roy B., Bouyssou, D. (1991) - Decision-Aid: an elementary introduction with emphasis on multiple criteria, LAMSADE, Cahier N. 106, Paris.

Roy B., Bouyssou, D. (1993) - Aide multicritère á la décision. Methodes et cas, Economica, Paris.

Rubin D.C. (1979) - On measuring fuzziness: A comment on "A fuzzy set approach to modifiers and vagueness in natural language", Jour. of experimental psychology: General, 108, pp. 486-489.

Saaty T.L. (1980) - The analytic hierarchy process, McGraw Hill, New York.

Sagoff M. (1988) - The economy of the earth, Cambridge University Press, Cambridge.

Samuelson P.A. (1942) - Constancy of the marginal utility of income, in Lange O., McIntyre F., Yntema T.O. (eds.)- Studies in mathematical economics and econometrics in memory of Henry Schultz, University of Chicago Press, Chicago.

Schärlig A. (1985) - Décider sur plusieurs critéres, Presses polytechniques et universitaires romandes, Lausanne.

Scitowsky T. (1941) - A note on welfare propositions in economics, Review of Economic Studies, vol. 9.

Scitowsky T. (1954) - Two concepts of external economies, Journal of Political Economy, vol. 58, pp. 143-151.

Scitovski T. (1976) -The joyless economy, Oxford University Press, New York.

Shackle G.L.S. (1967) - The years of high theory, Cambridge Press, Cambridge.

Sen A.K. (1961) - Optimising the rate of saving, Economic Journal, vol. 71, pp. 479-495.

Sen A.K. (1967) - Isolation, Insurance and the optimal rate of discount, Quarterly Journal of Economics, vol. 81, pp. 112-124.

Sen A. K. (1982) - Choice. Welfare and Measurement, Basil Blackwell, Oxford.

Shannon C.E. (1948) - A mathematical theory of communication, Bull. Syst. Tech., n 27.

Shapley L.S., Shubik M. (1954) - A method for evaluating the distribution of power in a committee system, American Political Science Review, 48, pp. 787-792.

Shubik M. (1983) - Game theory in the social sciences, the MIT Press, Cambridge.

Simon H. A. (1983)- Reason in human affairs, Stanford University Press.

Slowinski R. (1986) - A multicriteria fuzzy programming method for water supply systems development planning, Fuzzy Sets and Systems 19, pp. 217-228.

Slowinski R. (1990) - "FLIP": an interactive method for multiobjective linear programming with fuzzy coefficients, in Slowinski R., Teghem J.- Stochastic versus fuzzy approaches to multiobjective mathematical programming under uncertainty, Kluwer Academic Publishers, Dordrecht, pp. 249-262.

Slowinski R., Teghem J. (1990)- Stochastic versus fuzzy approaches to multiobjective mathematical programming under uncertainty, Kluwer Academic Publishers, Dordrecht.

Smith V.K. (ed.) (1979) - Scarcity and growth reconsidered, The Johns Hopkins University Press, Baltimore.

Smithson M. (1987) - Fuzzy set analysis for behavioural and social sciences, Springer-Verlag, New York.

Söderbaum P. (1992) - Neoclassical and institutional approaches to development and the environment, Ecological Economics 5 , pp. 127-144.

Solow R.M. (1974a) -The economics of resources or the resources of economics, American Economic Review, 64, pp. 1-14.

Solow R. M. (1974b) - Intergenerational equity and exhaustible resources, Review of Economic Studies, vol. 67, pp. 29-45.

Solow R. M. (1986) - On the intergenerational allocation of natural resources, Scandinavian Journal of Economics, vol. 88(1), pp.141-149.

Spronk J. (1981) - Interactive multiple goal programming for capital budgeting and financial planning, Martinus Nijhoff, Boston.

Steuer R.E. (1986) - Multiple criteria optimization- Wiley, New York.

Stewart T.J. (1991) - Decision analysis in public policy evaluation, paper presented at EURO XI, Aachen, Germany.

Stiglitz J.E. (1979) - A neoclassical analysis of the economics of natural resources, in Smith V.K. (ed.)- Scarcity and growth reconsidered, The Johns Hopkins University Press, Baltimore,, pp. 37- 65.

Stiglitz J.E. (1982) - The rate of discount for benefit-cost analysis and the theory of second best, in Lind R. (ed.) (1982) - Discounting for time and risk in energy policy, Resource for the Future, Washingtonpp, 151-204 .

Tanaka H., Asai K. (1984) - Fuzzy linear programming with fuzzy numbers, Fuzzy Sets and Systems 13, pp.1-10.

Tanino T. (1988) - Fuzzy preference relations in group decision making, in Kacprzyk J., Roubens M.- Nonconventional preference relations in decision making, Springer-Verlag, Berlin, pp. 54, 71.

Theil H.R. (1967) - Economics and information theory, North-Holland, Amsterdam.

Thevenet K., Rietveld P. (1992) - Sensitivity analysis with interdependent criteria for multicriteria decision making, Free University, Amsterdam, unpublished paper.

Thole U., Zimmermann H.J. and Zysno P. (1979) - On the suitability of minimum and product operators for the intersection of fuzzy sets, Fuzzy sets and systems, 2, pp. 167-180.

Thurstone L.L. (1969) - Attitudes can be measured, in Summers G.F. (ed.)- Attitude measurement, Chicago, Rand McNally.

Tinbergen J. (1956) - Economic policy: principles and design, North-Holland, Amsterdam.

Tomasin A. (1990) - L'ipotesi di parco del delta del Po. Materiali di analisi, Cassa di Risparmio di Padova e Rovigo, Padova (in Italian).

Tong R.M., Bonissone P.P. (1984) - Linguistic solutions to fuzzy decision problems, in Zimmermann H.J., Zadeh L.A. and Gaines B.R. (eds.)- Fuzzy sets and decision analysis, North-Holland, Amsterdam, pp. 323-334.

Turban E. (1990) - Decision support and expert systems- Macmillan Publ. Co., New York.

Turner R.K., Pearce D.W., Bateman I. (1994) - Environmental economics: an elementary introduction, Harvester Wheatsheaf, London.

Tversky A., Kahneman D. (1975) - Judgement under uncertainty: heuristics and biases, Science, vol. 185, pp. 1124-1131.

Vanderpooten D. (1989) - The interactive approach in MCDA: a technical framework and some basic conceptions, in Models and methods in multiple criteria decision making, edited by Colson G., De Bruyn Chr., Pergamon Press, Oxford, pp. 1213-1220.

Vanderpooten D., Vincke Ph. (1989) - Description and analysis of some representative interactive multicriteria procedures, in Models and methods in multiple criteria decision making, edited by Colson G., De Bruyn Chr., Pergamon Press, Oxford, pp. 1221-1238.

Vansnick J.C. (1986) - On the problem of weights in multiple criteria decision making (the noncompensatory approach), European Journal of Operational Research, 24 pp.288-294.

Vansnick J. C. (1990) - Measurement theory and decision aid- in Bana e Costa C.A. (ed.)- Readings in multiple criteria decision aid, Springer-Verlag, Berlin, pp. 81-100.

Verhoef E. (1993) - A partial equilibrium analysis of external costs: the compatibility of optimization, compensation and internalization, unpublished paper, Free University, Amsterdam, .

Victor P.A. (1991) - Indicators of sustainable development: some lessons from capital theory, Ecological Economics 4, pp. 191- 213.

Vincke Ph. (1985) - Multiattribute utility theory as a basic approach, in Fandel G., Matarazzo B., Spronk J. (eds.) Multiple criteria decision methods and applications, Springer-Verlag, Berlin, pp.27-40.

Vincke Ph. (1989) - L'aide multicritere à la decision- editions de l'Université de Bruxelles.

Vincke Ph. (1992) - Multicriteria decision aid, Wiley, New York.

Voogd H (1983) - Multicriteria evaluation for urban and regional planning, London:Pion.

Wallace R.R., Norton B.G. (1992) - Policy implications of Gaian theory, Ecological Economics 6, pp. 103-118.

Wallsten T.S., Budescu, D.V., Rapoport A., Zwick R. and Forsyth B. (1986) - Measuring the vague meaning of probability terms, Jour. of Experimental psychology: General, 115, pp. 348-365.

Wierzbicki A.P. (1982) - A mathematical basis for satisficing decision making, Mathematical Modelling, 3, pp. 391-405.

Williams B. (1979) - Internal and external reasons, in Harrison R.- Rational action, Cambridge University Press, pp. 17-28.

Winkler R.L., Hays W.L. (1975) - Statistics, Holt, Rinehart and Winston, New York.

World Commission on Environment and Development (1987) -Our common future, Oxford University Press, Oxford.

Yager R.R. (1978) - Fuzzy decision making including unequal objectives, Fuzzy Sets and Systems, 1 pp. 87-95,.

Yager R.R.(1979) - On the measure of fuzziness and negation part I: membership in the unit interval, International Journal of General Systems, 5, pp. 221-229.

Yu P.L. (1973) - A class of solutions for group decision problems, Management Science, vol. 19, N. 8, pp.936-946.

Yu P.L. (1985) - Multi criteria decision making: concepts, techniques and extensions, Plenum Press, New York.

Yu P.L. (1990) - Forming winning strategies: an integrated theory of habitual domains, Springer-Verlag, Berlin.

Zadeh L.A. (1965) - Fuzzy sets, Information and Control, 8, pp. 338-353.

Zadeh L.A. (1972) - A fuzzy set theoretic interpretation of linguistic hedges, Journal of Cybernetics, 2, pp. 4-34.

Zadeh L.A. (1973) - Outline of a new approach to the analysis of complex systems and decision processes, IEEE Trans., Vol. SMC 3, pp.28-44.

Zadeh L.A., Fu K. S., Tanaka K. and Shimura M. (eds.) (1975) - Fuzzy sets and their applications to cognitive and decision processes, Academic Press, New York.

Zadeh L.A. (1983) - A computational approach to fuzzy quantifiers in natural languages, in Comp. and Maths. with Appls., 9, pp. 149-184.

Zeleny M. (1974) - A concept of compromise solutions and the method of the displaced ideal, Computers and Operations Research, vol. 1, N. 4, pp. 479- 496.

Zeleny M. (1982) - Multiple criteria decision making, McGraw Hill, New York.

Zétényi T. (1988) - Fuzzy sets in psychology, North-Holland, Amsterdam.

Zimmermann H.J. (1978) - Fuzzy programming and linear programming with several objective functions, Fuzzy Sets and Systems, 1, pp. 45-55.

Zimmermann H. J. (1980) - Testability and meaning of mathematical models in social sciences, in Mathematical modelling, Vol.I pp. 123-139, Pergamon Press, (USA).

Zimmermann H.J., Zysno P. (1983) - Decisions and evaluations by hierarchical aggregation of information, Fuzzy Sets and Systems 10 pp.243-260.

Zimmermann H.J. (1986) -Fuzzy set theory and its applications, Kluwer-Nijhoff Publishing, Boston.

Zimmermann H.J. (1987) -Fuzzy sets, decision making, and expert systems, Kluwer-Nijhoff Publishing, Boston.

Zionts S. (1982) - Multiple criteria decision making: an overview and several approaches, State University of New York at Buffalo, Working Paper No. 454.

Zwick R., Budescu D.V. and Wallsten T.S. (1988) - An empirical study of the integration of linguistic probabilities, in Zétényi T.- Fuzzy sets in psychology, North-Holland, Amsterdam, pp. 91-125.

Springer-Verlag
and the Environment

We at Springer-Verlag firmly believe that an international science publisher has a special obligation to the environment, and our corporate policies consistently reflect this conviction.

We also expect our business partners – paper mills, printers, packaging manufacturers, etc. – to commit themselves to using environmentally friendly materials and production processes.

The paper in this book is made from low- or no-chlorine pulp and is acid free, in conformance with international standards for paper permanency.